MW00720447

DESIGNS IN SCIENCE

MOVEMENT

SALLY and ADRIAN MORGAN

Facts On File, Inc.
460 Park Avenue South
New York NY 10016

Library of Congress Cataloging-in-Publication Data
Morgan, Sally.
MOVEMENT / Sally and Adrian Morgan.
p. cm. — (Designs in science)
Includes index.
Summary: Discusses movement, its importance for survival of plants, humans, and other animals, and the various ways in which it is achieved.
ISBN 0-8160-2979-2
1. Motion — Juvenile literature. 2. Mechanical movements—Juvenile literature. 3. Human mechanics —Juvenile literature [1. Motion.]
I. Morgan, Adrian. II. Title. III. Series: Morgan, Sally. Designs in science.
QC133.5.M67 1993
531. 1'1—dc20 93-20162

Facts On File books are available at special discounts when purchased in bulk quantities for businesses, associations, institutions or sales promotions. Please call our Special Sales Department in New York at 212/683-2244 or 800/322-8755.

10 9 8 7 6 5 4 3 2 1

This book is printed on acid-free paper.

Editor: Su Swallow
Designer: Neil Sayer
Production: Peter Thompson
Illustrations: Hardlines, Charlbury
David McAllister

Acknowledgments

For permission to reproduce copyright material the authors and publishers gratefully acknowledge the following:

Cover Michael Leach, Oxford Scientific Films
Title page Dayle Boyer, NASA/Science Photo Library **Contents page** Sally Morgan, Ecoscene **page 4** (top) Robert Harding Picture Library (bottom) Sally Morgan, Ecoscene **Page 5** Anthony Bannister, NHPA **page 6** Anthony Cooper, Ecoscene **page** 7 (top) Sally Morgan, Ecoscene (bottom) Peter Parks, Oxford Scientific Films **page 8** (top) Ken Lucas, Planet Earth Pictures (bottom) Michael Melford, The Image Bank **page 9** Sally Morgan, Ecoscene **page 10** Sally Morgan, Ecoscene **page 11** (top) Peter Scoones, Planet Earth Pictures (bottom) Howard Hall, Oxford Scientific Films **Page 12** Stephen Dalton, Science Photo Library **page 13** (top) Winkley, Ecoscene (middle) Rout, Ecoscene (bottom) Andrew Mounter, Planet Earth Pictures **page 14** Andy Callow, NHPA **page 15** Alastair Macewen, Oxford Scientific Films **page 16** (top) Dayle Boyer, NASA/ Science Photo Library (bottom and page 17) Stephen Dalton, NHPA **page 18** Stephen Dalton, NHPA **page 19** (left) Robert Harding Picture Library (right) Ben Osborne, Oxford Scientific Films **page 21** (top) Ken Lucas, Planet Earth Pictures (bottom) Manfred Danegger, NHPA **page 22** (top) Robert Harding Picture Library (bottom) Stephen Dalton, NHPA **page 23** (top) Schröter/Bildagentur Schuster/Robert Harding Picture Library (bottom) Sally Morgan, Ecoscene **page 24** (top) Gordon Maclean, Oxford Scientific Films (bottom left) Sally Morgan, Ecoscene (bottom right) Robert Francis, Robert Harding Picture Library **page 25** Stephen Dalton, NHPA **page 26** (left) Stephen Dalton, NHPA (right) Robert Harding Picture Library **page 27** (left) Robert Harding Picture Library (right) John Sanford, Science Photo Library **page 28** (top) Kevin Burchett, Bruce Coleman Ltd (middle) Lester, Ecoscene (bottom) Peter Menzel, Science Photo Library **page 29** (top) Dr David Jones, Science Photo Library (middle) Science Photo Library (bottom) Michael Klinec, Bruce Coleman Ltd **page 30** Sally Morgan, Ecoscene **page 31** TKM Automotive Ltd **page 32** (top) Gérard Lacz, NHPA (bottom) Stephen Dalton, NHPA **page 33** (top) Stephen Dalton, NHPA (bottom) Towse, Ecoscene **page 34** Hawkes, Ecoscene **page 35** Mike Devlin, Science Photo Library **page 36** Robert Harding Picture Library **page 37** (top) John Hayward, NHPA (bottom) Anthony Bannister, NHPA **page 38** (top) Stephen Dalton, NHPA (bottom) Towse, Ecoscene **page 40** Sally Morgan, Ecoscene **page 41** (top) Cooper, Ecoscene (bottom) Sally Morgan, Ecoscene **page 42** (top) Norbert Wu, Oxford Scientific Films (middle) Ron Church, Science Photo Library (bottom) Lockheed, Aviation Picture Library **page 43** Alex Bartel, Science Photo Library

DESIGNS IN SCIENCE

MOVEMENT

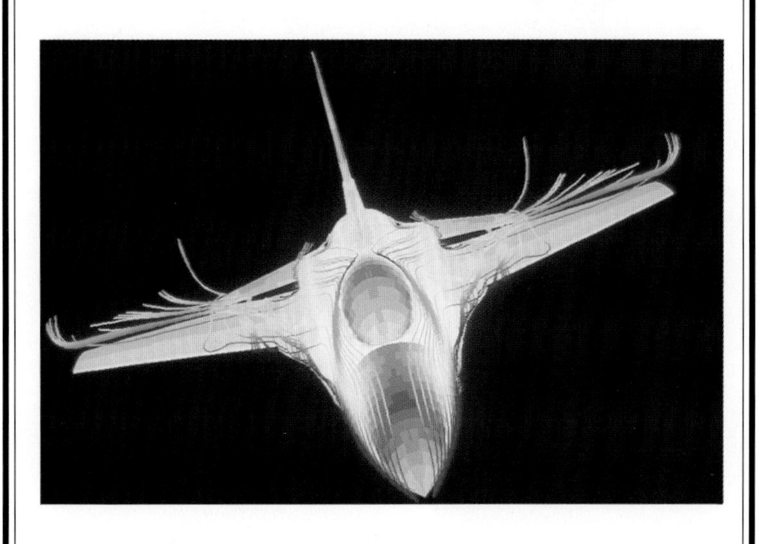

SALLY and ADRIAN MORGAN

Facts On File

NOTE ON MEASUREMENTS:

In this book, we have provided U.S. equivalents for metric measurements when appropriate for readers who are more familiar with these units. However, as most scientific formulas are calculated in metric units, metric units are given first and are used alone in formulas.

Measurement

These abbreviations are used in this book.

METRIC		U.S. EQUIVALENT	
	Units of length		
km	kilometer	mi.	mile
m	meter	yd.	yard
cm	centimeter	ft.	foot
mm	millimeter	in. or "	inch
	Units of volume		
l	liter	gal.	gallon
ml	milliliter	qt.	quart
		pt.	pint
m^3	cubic meter	cu. ft.	cubic foot
cm^3	cubic centimeter	cu. in.	cubic inch
	Units of mass		
g	gram	oz.	ounce
kg	kilogram	lb.	pound
	Units of temperature		
°C	degrees Celsius	°F	degrees Fahrenheit

Movement is one book in the seven-volume series Designs in Science. The series is designed to develop young people's knowledge and understanding of the basic principles of movement, structures, energy, light, sound, materials, and water, using an integrated science approach. A central theme running through the series is the close link between design in the natural world and design in modern technology.

Contents

The Concorde, one of the most advanced passenger planes, flies faster than the speed of sound.

The fastest speed recorded for a bird is over 290 km/h (181 mph), by a peregrine falcon that was swooping. Spine-tailed swifts reach speeds of 170 km/h (106 mph).

Divers use flippers to propel them through the water. Fish use their tails for forward movement.

Introduction

Birds and airplanes can fly through the air. Fish and submarines can move through water. Cats and cars can move fast on land. These are only a few examples of different forms of movement.

Over millions of years nature has evolved many designs to help animals achieve movement. Over a much shorter period, people have designed machines that can move. By looking at the designs evolved by nature and those produced by people, we can gain a good understanding of how movement is achieved. Design engineers themselves are finding the solutions to some of their problems by looking at the natural world.

All animals and machines have to obey the same laws of physics. It should not surprise us to find that animal construction is often very similar to that of machines when animals and machines are carrying out the same functions. For example, some of the most modern aircraft use specially shaped flaps at the tips of their wings to improve performance, but birds have evolved the same wing shapes naturally over millions of years.

Why is movement necessary at all? Animals have to move in order to find food, to find homes and to escape from danger. Plants also depend on movement for survival. Although they do not "get up and walk," they use air and water currents to spread their seeds, and water and food move through tubes in their stems to reach the leaves.

How is movement produced? An object will stay in one place until a force is applied to it. In other words, to move an object, work has to be done. Energy is the ability to do work. There are many forms of energy — movement energy, electrical energy, chemical and heat energy. Movement often involves converting one form of energy into another. A car converts the chemical energy in gas into heat, and then into movement energy. When you run, you convert chemical energy from your food into movement energy.

Energy conservation is very important in design because the conversion of energy from one form to another is never 100 percent efficient. Car engines would be much more efficient if they could turn all the chemical energy in gas into movement energy. As it is, the heat energy created is simply pushed out into the atmosphere and wasted. When designing anything that moves, it is important to minimize such losses.

Moving on land, in air and in water

> **!** *The blue whale can grow to 30 m (99 ft.) in length and weigh over 160 tonnes (176 tons). The largest dinosaur, brachiosaurus, only weighed about 80 tonnes (88 tons).*

Movement on land is simplified by the fact that the ground is solid. It supports weight and allows animals and machines to push against it, making it relatively easy for them to move across the ground. Air is a gas and water is a liquid, however, which creates problems for animals and machines moving through them. There is no solid ground to support their weight, and they meet less resistance when they push against air and water.

Swimming and flying use very similar principles. Fish do not swim as fast as birds can fly because water is harder to move through, but water provides more support, which makes it a more suitable medium for large animals to inhabit. It is not surprising, therefore, that the world's fastest animals are found in the air, and the largest live in the oceans. Many heavy objects seem to weigh less in water than on land. Fishermen use a landing net for this very reason. When the fish is under water, the rod and line are strong enough to hold the weight of the fish, but out of water the fish is no longer supported by the water and the rod or line may break, so the net is used to lift the fish onto the bank.

An object moving through air or water will experience some resistance to movement — the molecules of the air or water have to be pushed out of the way. This resistance is called drag. Drag affects animals and machines moving in water and through the air, but it has the most effect in water because water is denser than air. Each group of animals experiences drag but solves the problem in different ways. These ways will be discussed in the following chapters.

This book looks at how animals and machines move through water, in the air and on land. It also looks at the movement of fluids, in animals, plants and machines, and looks forward to designs for movement in the future.

Important words are explained at the end of each section, under the heading **Key words**, and in the glossary on page 46. You will find some amazing facts in each section, together with some experiments for you to try and questions for you to think about.

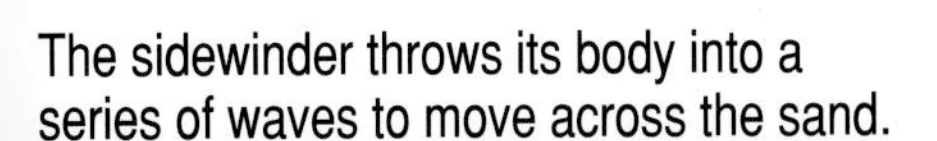

The sidewinder throws its body into a series of waves to move across the sand.

Key words

Drag the resistance to movement. The flow of air or water over a moving object tends to slow it down.

Resistance any force that slows down or opposes movement.

Moving in water

Nearly two thirds of our planet is covered by water. Life began in the oceans more than 3,500 million years ago. Today, many different types of animals can be found living in water, from the tiniest single-celled bacteria and algae to the largest living animal, the blue whale.

Critical factors that affect the ability of animals and machines to move on and underneath the surface of the water are whether they float or sink and how well they are streamlined.

Floating and sinking

The water of the Dead Sea is so dense that you can float sitting up in it.

Why does a steel ship float on the surface of water while a solid lump of steel of the same mass sinks?

Floating and sinking involve density as well as the mass of an object. If the density of an object is more than the density of water it will sink; if it is less then it will float. The steel ship floats because it contains a lot of air, so it is less dense than the water, while the solid lump of steel sinks because it is more dense than the water. Solids and liquids have a higher density than gases because their atoms are closely packed together.

An object's density is calculated by dividing its mass by its volume.

$$\textbf{density} = \frac{\textbf{mass}}{\textbf{volume}}$$

A solid lump of silver with a mass of 200 g and a volume of 20 cm^3 has a density of 10 g/cm^3. A hollow tube of silver with a mass of 200 g and a volume of 100 cm^3 has a density of 2 g/cm^3.

Ships are carefully loaded so that they do not sink too deeply in the water.

When something is lowered into water, the water that surrounds it pushes up on it. This is called upthrust. If the upthrust is greater than the weight of the object itself, then the object will float, and we say it is buoyant. If the upthrust is less than the weight of the object, the object sinks. If the upthrust and weight of the object are equal, the object will stay at one level in the water, and we say it has neutral buoyancy. A ship is designed so that it can carry cargo and still float. Large ships have Plimsoll lines on the hull to indicate the normal water level, and the safe level of loading. There are different levels for winter and summer, fresh and salt water, because the density of sea water itself can vary with the temperature of the water and the amount of salt present.

Ships float at different levels in salt and fresh water. Will a ship float higher in salt or fresh water? How could you check your answer?

EXPERIMENT

A simple hydrometer

One way of comparing the density of fluids is to use a simple hydrometer. You will need a drinking straw, some Plasticine (oil-based modeling clay), a large glass beaker or jam jar, some water, milk and salt.

1 Make a simple hydrometer from a 10 cm (4") length of a drinking straw with a knob of Plasticine at one end.

2 Place the hydrometer in a beaker of water and check that the hydrometer floats upright. Adjust the amount of Plasticine if necessary.

3 Mark the straw with a line level with the surface of the water.

4 Remove the hydrometer and add 10 cm^3 (.3 fluid ounces) salt to the water, making sure the salt dissolves completely. Put the hydrometer back in the salt water and make a new pencil line level with the surface of the water. Look at the two lines. What effect does the salt have?

5 Repeat the experiment, this time comparing ice cold water and warm water.

6 Repeat using other fluids, such as milk, and compare the levels.

Depth control

Animals that live in water have a particular problem — they need to be able to control their depth. There are two methods of achieving this. For short periods, some animals can stay at depth by using fins or flippers. When they stop moving, they rise up in the water. This method is best suited to animals that are naturally buoyant, and that need to surface from time to time anyway in order to breathe, such as whales and seals. However, controlling depth with fins or flippers uses up valuable energy.

For animals that need to spend long periods underwater, often at considerable depth, it is much more efficient to be neutrally buoyant. This avoids the need to use energy just to remain at the required depth, and so the majority of marine creatures have evolved ways of maintaining neutral buoyancy.

Many of the smaller organisms, such as plankton, have lots of appendages, which give them a large surface area. This large surface area means they sink very slowly through the water, and they can easily stay at one level by making small movements. Deposits of fat and oil, which are both less dense than water, also help to make some planktonic animals and plants buoyant.

Sharks have large fatty livers to aid buoyancy. A shark's liver makes up almost 20 percent of its entire body weight. Even so, sharks are denser than water so they have to swim continuously in order to avoid sinking. They have a pair of fins that act like small wings and give a little lift and so aid buoyancy.

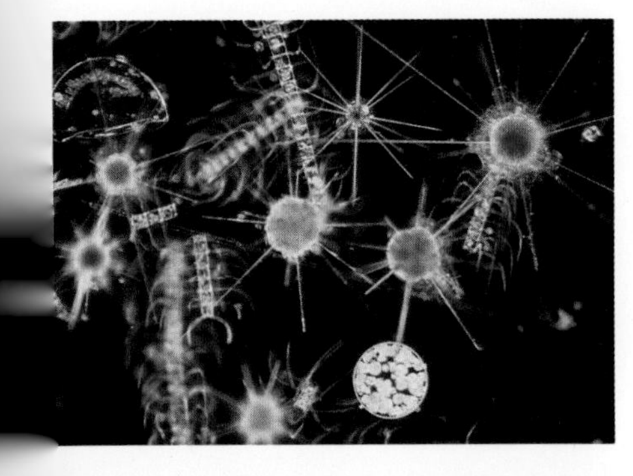

There are thousands of different types of plankton, but they all have a large surface area.

!

If deep-swimming fish are brought to the surface too quickly, as sometimes happens when they are caught by fishing boats, the rapid decrease in outside pressure causes the swim bladder to expand and come out through their mouth.

Bony fish have evolved a unique organ called a swim bladder, which is a small air sac in the abdominal cavity near the backbone. The amount of air inside the swim bladder can be controlled to alter the fish's buoyancy. It is easy to see how goldfish adjust the amount of air in their swim bladder. They swim to the surface to gulp air in, to keep high in the water, and blow out bubbles of air in order to sink. In other fish, the swim bladder is sealed. A special gland transfers oxygen between the swim bladder and the blood, automatically adjusting the air in the swim bladder to help the fish change depth.

Wax is an unusual substance, since its density depends upon whether it is in a solid or liquid state. In solid form it takes up less space, and so increases in density. The sperm whale has wax in its body, and controls the temperature of the wax to help it surface and dive. It has a very large head that contains an organ filled with wax. As the whale inhales cold water, the wax cools and shrinks, and so does the head of the whale. As a result, the overall density of the whale increases, which helps the whale to dive. To ascend, the whale exhales the cold water, and allows its normal body temperature to warm up the wax. The wax becomes less dense and the whale becomes more buoyant.

The design of a submarine imitates some of the designs of the underwater animal kingdom. Submarines have ballast tanks, which can either be pumped full of air or be opened to the sea to

The nautilus (below) has a series of gas-filled chambers inside its shell. It alters its buoyancy by changing the amount of gas. The submarine (bottom) uses the same method to control its depth.

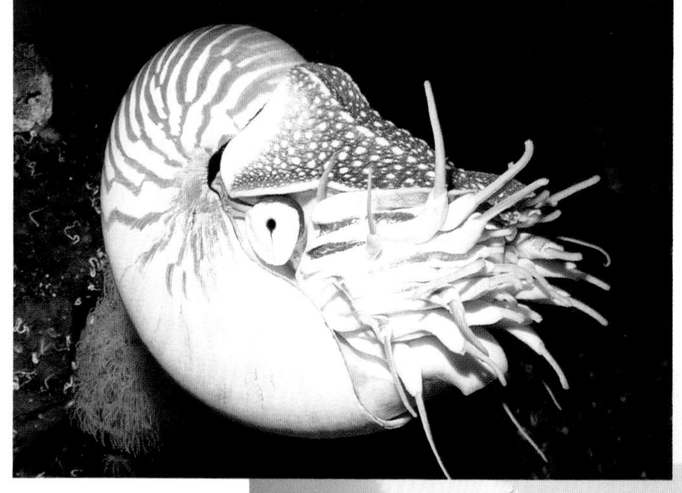

What similarities can you think of between the swim bladder of the goldfish and the ballast tanks of a submarine?

fill with water. When full of water, the submarine sinks. If compressed air is blown into the tanks, the density of the submarine becomes less than that of the surrounding water, and the submarine rises. The angle of ascent or dive depends on which of the many ballast tanks are filled with air or water. In an emergency, all the tanks are filled with air at once, to make the submarine surface as fast as possible. Extra control is gained by changing the angle of the hydroplanes, which are small wing-shaped fins at the fore and aft ends of the hull.

EXPERIMENT

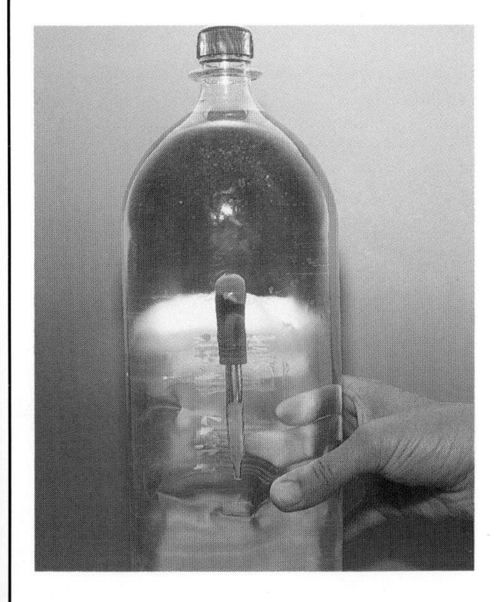

Buoyancy

This experiment investigates the effect of pressure on the buoyancy of an object. You will need a 2 liter clear plastic bottle with cap, an eye dropper, and a tall glass beaker.

1 Fill the glass beaker almost to the brim with water.

2 Fill the eye dropper with water to the point that it will float in water with its top only just above the surface.

3. Fill the bottle with water, and then carefully transfer the dropper to the bottle without losing any water out of it. It is important that the volume of water in the dropper stay the same.

4 Screw the bottle cap in place tightly.

5 Squeeze the bottle.

What happens to the dropper? How can you make it rise and sink?

Explanation Squeezing the bottle increases the water pressure inside. The air in the dropper cannot escape, so it is compressed and more water enters. The dropper has less air, becomes denser and less buoyant, so it sinks. By releasing the pressure, air expands and displaces the water out of the dropper, so it rises.

Moving forward

Fish achieve forward movement by alternately contracting and relaxing sets of muscles on either side of the backbone. A series of waves moves along the body, moving the tail and tail fin from side to side, which propels the fish forward. This type of movement can be clearly seen in eels. Dolphins and whales move their tail up and down, but the principle is identical.

Ships use propellers to produce forward movement. Most large ships have a propeller with at least four blades. The propeller blades work like screws to propel the water backward and so cause the ship to move forward. The power to drive big ships comes from steam turbines. Steam is produced in huge boilers and is piped into the turbine. Jets of steam force the blades of the turbine to turn and spin a shaft. The shaft is attached to the propeller through a transmission, so the spinning of the shaft causes the propeller to turn.

! *The world's first turbine-driven vessel, the British* Turbinia, *appeared in 1897. It reached speeds of more than 30 knots (about 60 km/h or 37.5 mph).*

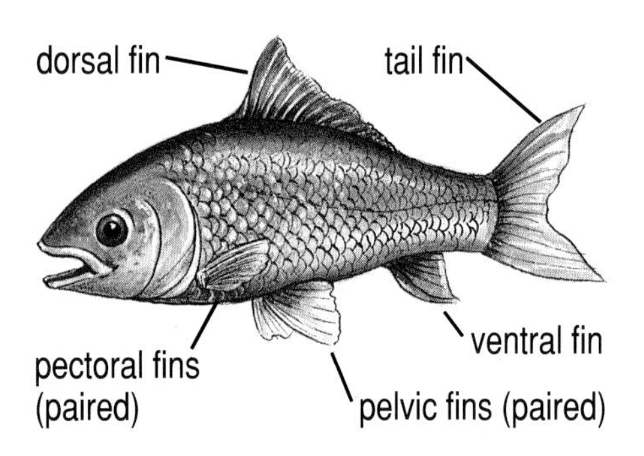

Fish use their tail fin for forward movement, and other fins to control their movement. The dorsal and ventral fins aid stability, while pectoral and pelvic fins are used for steering and balance. The angle of the fins allows fine control over turning, climbing, diving or helping the fish stay at a constant depth. On a larger scale, ships have fins known as stabilizers, fitted to the side of the hull below the waterline. They are used to reduce roll, particularly in bad weather.

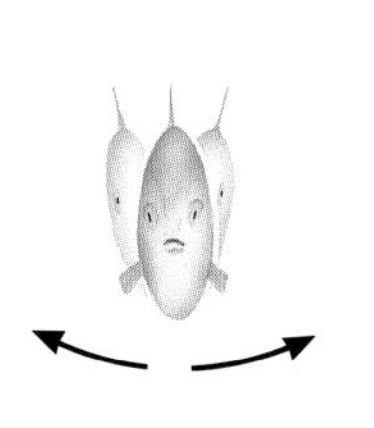

The dorsal and ventral fins prevent rolling.

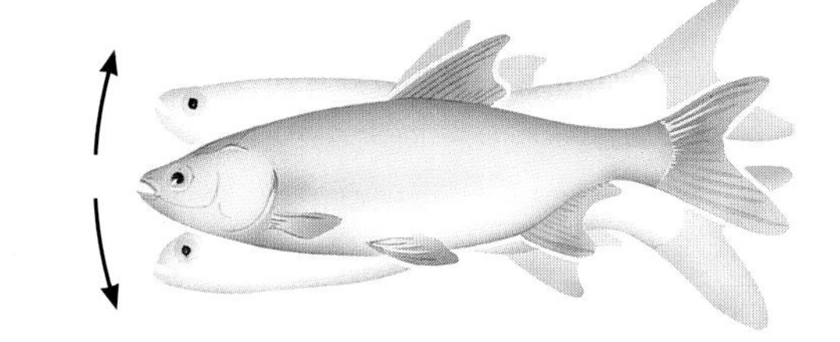

The paired fins (pectoral and pelvic) control up-and-down movement.

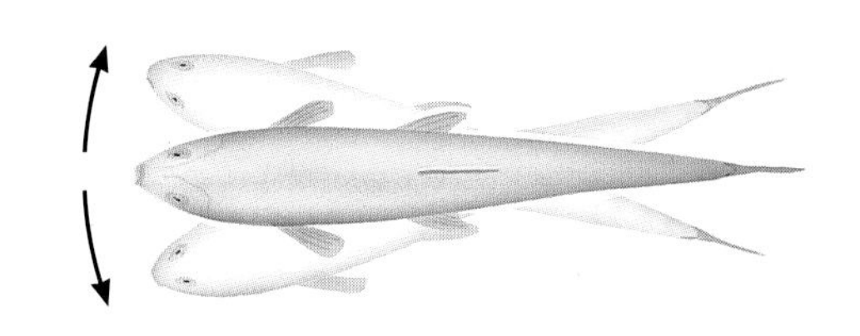

The dorsal and ventral fins also prevent yawing.

Streamlined shapes

As an animal or machine moves through water it experiences drag (see page 7). As the speed increases, so does the drag. But drag increases by a greater amount. If a fish doubles the speed at which it swims, it has to overcome four times the drag. It is therefore vitally important to almost all marine animals, and particularly the larger ones, to keep drag to an absolute minimum. Fish and marine mammals have evolved many different techniques to minimize drag and maximize power output.

EXPERIMENT

Streamlining

This experiment enables you to work out the best shapes for streamlining. You will need a tall glass measuring cylinder or storage jar (the type used for pasta), a little wallpaper paste, Plasticine and a stopwatch.

1 Fill the measuring cylinder or storage jar with wallpaper paste. This sticky liquid will slow down the movement of objects so that you will be able to time their descent.

2 Divide the Plasticine into pieces that weigh approximately 20 g (.7oz.) and make up a range of shapes such as a sphere, a cube, a cylinder and tear drop.

3 Using a stopwatch, record the time it takes for each object to fall from the top to the bottom of the cylinder. The object that is most streamlined will have the fastest time.

Which shape was the most streamlined? Which was the least streamlined?

? *How many similarities can you think of between a shark, a dolphin and a torpedo?*

One way to minimize drag is to have a smooth, slippery shape — this is called streamlining. It is particularly important to animals and machines in water as well as to fast-moving airborne animals and machines.

When water flows around an object, it can flow in one of two ways. When the flow is smooth, it is called laminar, and when it is rough, it is called turbulent.

Laminar flow causes less drag than turbulent flow. Animals that move fast create little turbulence.

Fish are tapered at each end, and have smooth surfaces, so that water flows easily around them. The scales of a fish are moistened by a slimy mucus to reduce surface friction. Fins are essential for controlling movement, but they stick out from the body and create extra drag. Some fish have evolved smaller fins to enable them to move fast through the water. Predatory fish such as barracuda, tuna, billfish and swordfish have very streamlined body shapes and have almost reached swimming perfection. They have very pointed snouts, some of which end in a long spike, and the body tapers gently, ending in a crescent-shaped tail. There is no bulge that might interrupt laminar flow. Even their eyes fit flush with their body surfaces. The tuna has special scales just behind the head that act as a spoiler, producing a slight turbulence around the widest part of the body, and thereby reducing drag over the tail end. When swimming, its control fins slot into special grooves to avoid drag.

The dolphin is a highly efficient swimmer, capable of great speed underwater. It is particularly well streamlined: the widest part of its body is set far back toward its tail, so that there is the

Fish such as the barracuda (below) and mammals such as the dolphin (bottom) solve the problems of moving fast through the water in the same way. They have both evolved streamlined shapes.

Spoilers on sports cars are not used to reduce drag. Can you think why they are fitted?

maximum laminar flow over the front of its body before turbulent flow appears. This is not the only technique employed to reduce drag, however. The dolphin is able to get a layer of water with laminar flow to stick to its skin surface. Its skin is soft and at the point where laminar flow starts to break away, the skin is deformed into tiny ridges. This sets up very small areas of local turbulence between each ridge. Each tiny turbulent area lowers the water pressure, and causes the laminar flow to stick to the skin a little longer. This delays the start of large-scale turbulent flow.

Some of these design features for reducing drag are used by people. Many modern jet aircraft have projections on the upper surfaces of their wings, which create micro-turbulence and delay the break up of laminar flow. Golf balls have tiny dimples in their surface. This creates micro-turbulence, enabling them to fly more easily through the air.

The sailfish holds the speed record for any marine animal, at 110 km/h (69 mph) over a short distance, faster than the land speed record of the cheetah, which is approximately 100 km/h (63 mph).

The surface of a golfball and the skin of a dolphin both give extra speed. Designers are now looking more carefully at the natural world for other ideas to improve their designs.

Skimming the surface

Some animals live in water and air and have to be able to move in both. Ducks, for example, have webbed feet for swimming and feathers for flying. Their feathers have the same structure as those of other birds, but they are also oiled to prevent waterlogging. The ducks' streamlined bodies are equally useful in the air and under the water.

Flying fish can take to the air, and glide for tens of yards, in order to escape predators. They propel themselves out of the

The hydrofoil (above) and the hovercraft (right) are both designed to move across the surface of the water.

water at high speed, and then use their large pectoral fins as wings. They sometimes land on boat decks, having crashed into the sails.

People have not yet developed a flying submarine, except in science fiction films. Nevertheless, hovercraft and hydrofoils skim over the surface of the water rather than pushing through the water. This reduces drag considerably, and they can reach speeds of up to 80 km/h (50 mph). A hydrofoil has small wings, fixed to legs under the hull. It moves off like a conventional boat, driven by a propeller, but as the speed increases the wings produce lift and the boat begins to rise. The propeller is mounted below the rear wing so that it always stays underwater.

The hovercraft works on a different principle. It rides on a cushion of air. Giant fans pump air underneath it, and a rubber skirt stops the air from escaping too fast. Even if the fans stop, the hovercraft is very safe, gently coming to rest on its body, which can float in water. Forward motion and directional control are provided by the propellers. Because the hovercraft rides on a cushion of air, it can operate as well on land as it can at sea.

The Portuguese man-of-war is a jellyfish that has taken buoyancy almost to the limit. It has evolved gas floats that allow it to float right on the surface of the water. It inflates the floats, which then catch the wind like sails and carry the jellyfish across the surface of the sea.

Sails were first used by humans over 5,000 years ago, by the Egyptians. Since 1783,

The float of this Portuguese man-of-war lies above the water to catch the wind. The float may be 30 cm long (12"), and stands 15 cm (6") above the water.

Sailing before the wind

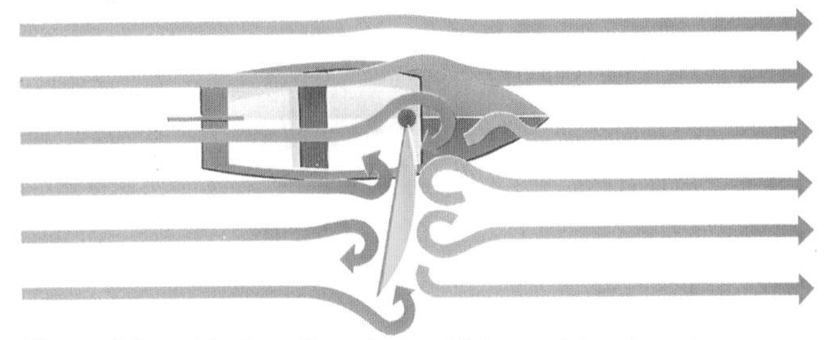

The sail is set to trap the wind, which provides thrust. Turbulence is created because the sail interrupts the airflow.

Sailing into the wind

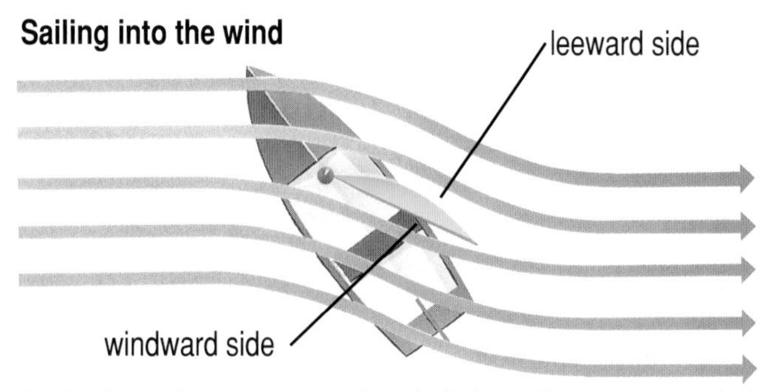

As the boat zigzags across the wind, the sail traps some wind on the windward side, which raises the air pressure. The air flows faster on the leeward side, creating an area of low pressure. The difference in pressure provides sailing thrust.

when the first steam-powered boat was built, sails have become less important, although even today dhows and other traditional boats are still used to carry cargo and passengers in Africa and the Far East.

The aim of a sail is to catch the wind. Bigger sails trap more wind and so produce more movement than smaller sails. A sailing boat can move in almost any direction, regardless of where the wind is blowing from, apart from directly into the wind. If the wind is behind the boat, the sail is held at right angles to the boat. The action of the wind on the sail thrusts the boat forward. If a sailing boat wants to sail directly toward the direction from which the wind is coming, it has to travel in zigzags. This is known as tacking.

In strong winds, sailing boats use smaller sails. Why do you think they do this?

Surface tension

Some small animals move and live on the very surface of the water, not by floating but by walking on the surface. These animals are actually denser than water, and so would normally sink. However, they are able to walk on the surface because of an effect called surface tension.

If you fill up a glass very slowly, the water will rise above the edge of the glass before it finally overflows. It looks as if there is a skin holding the water in place. This is known as surface tension. The bulge of water is held together by water's tendency to stick to itself. Eventually the pressure of water forming the bulge gets too much and water spills over the edge of the glass. Surface tension can also be seen on drops of water, such as raindrops or dewdrops on a spider's web.

Water molecules are attracted to each other and they form bonds. These bonds are quite difficult to break so the water holds together. At the surface of the water, molecules can only bond with those molecules at the sides and below them, and cannot bond with the air molecules, so they appear to form a tight skin.

Surface tension causes water droplets to form spheres when they come to rest on a surface.

This large raft spider is supported by surface tension as it lies in wait for its prey.

Surface tension is exploited by a number of animals. Such animals must be light so they do not break through the surface. Pond skaters make use of surface tension to move around on water. They can skim quickly across the surface of water looking for prey. They have wax-coated feet that are able to stand on the film of water without breaking it. Their six legs are splayed wide apart, each creating a tiny dimple on the surface. Surface tension is surprisingly strong, and it can support the weight of the raft spider, which can grow up to 10 cm (4") across. The legs of both pond skater and raft spider are very sensitive to tiny vibrations. This helps them to locate their prey, such as a fly that has fallen in the water.

? *What happens if you move a drop of water around on a flat surface, for example by drawing it along with your finger?*

Key words

Buoyancy the ability to rise or float in water.

Density a measure of how closely the atoms or molecules of a substance are packed together. It is calculated by dividing the mass of an object by its volume.

Streamlining the design of a smooth, slippery shape that keeps drag to a minimum.

Surface tension molecular force that pulls the surface of a liquid into the minimum area possible.

? *Why do you think a pond skater keeps its six legs splayed wide apart?*

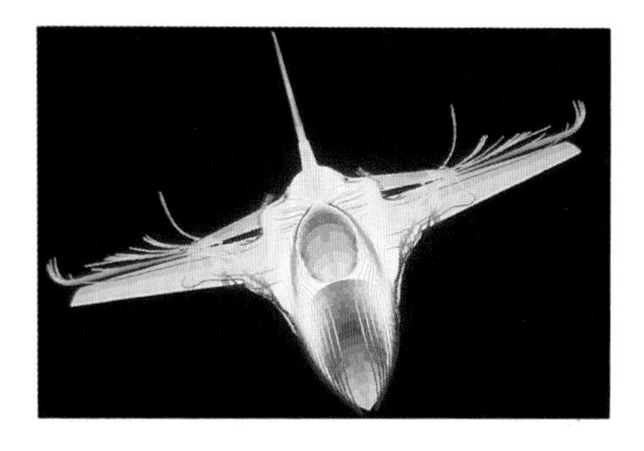

Taking to the air

The first animals to take to the air, hundreds of millions of years ago, were insects, followed by birds and bats. The first planes did not appear until the beginning of this century.

A computer image of airflow over a modern jet aircraft

Wing design

Airflow patterns

Air passing over the top of a wing has to travel further than the air passing below, so it moves faster. The faster air moves, the lower its pressure, so the air above the airfoil is at a lower pressure than the air below. The higher pressure below forces the airfoil up, which provides lift. The amount of lift depends on the angle at which the wing meets the air. The lift increases as the angle becomes steeper. However, if the angle is too steep the air flowing over the airfoil becomes turbulent. More turbulence means more drag, and this reduces speed. If too much lift is lost, the animal or aircraft will drop quickly. This is called stalling.

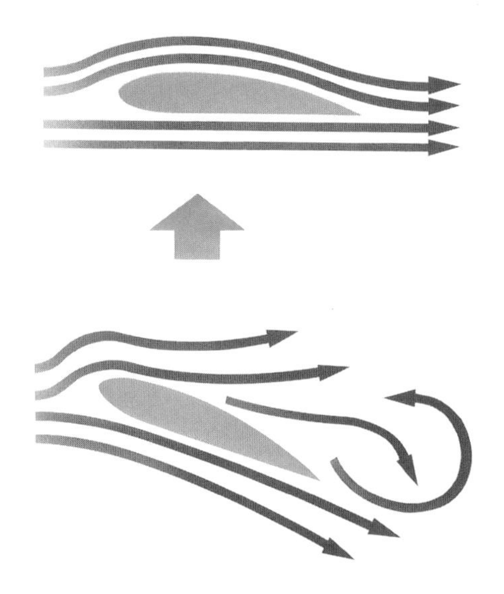

The design principles that apply to the wings of flying animals are the same as those that apply to modern aircraft. Anything that flies is acted upon by four forces:

1 Weight - force acting downward as a result of gravity.
2 Lift - the upward force that counteracts weight.
3 Drag - the resistance that slows down an object moving through air.
4 Thrust - the force that moves an object forward.

The key factor affecting flight is lift, which is produced by the wings. All wings have the same basic shape: an aerodynamic shape called an airfoil.

A little owl in flight. The down beat of the wings pushes the bird forward.

EXPERIMENT

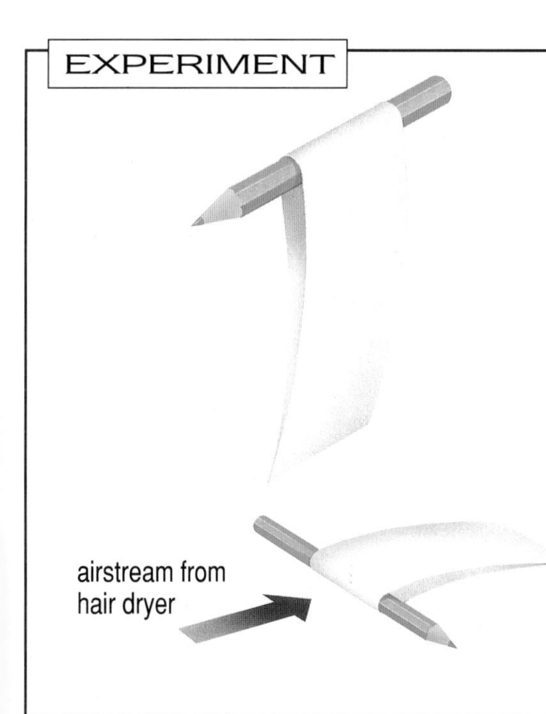

Design an airfoil

You will need some thin card stock, a pencil and a hair dryer.

1 Cut out a piece of stiff card approximately 20 cm x 5 cm (8" x 2").

2 Stick the two short ends of the card together as shown in the diagram.

3 Hang the wing over a pencil.

4 Produce an airstream using the hair dryer.

5 Place the wing in the airstream. The wing will take up a horizontal position.

What force is keeping it there? Move the hair dryer closer to the wing, then further away. What effect does increasing or decreasing the airstream have on the wing?

Some dinosaurs could fly. Pteranodon had a wingspan of over 8m (26 ft.), the size of a small plane.

First let us look at feathers, and the design of the bird wing. It is the wing that gives the bird its lift and thrust. There are several types of feathers, each with a particular function. The long primary feathers on the front of the wing are used mainly for propulsion, while the secondary feathers on the inner part of the wing give the wing its curved airfoil shape and provide most of the lift. Some birds have a tuft of feathers on the front edge. These decrease the stalling speed and help low-speed control, which is particularly useful when landing. The body feathers provide a smooth contour to reduce drag, while the tail feathers can alter the shape and angle of the tail and hence affect lift and provide extra control.

When a bird flaps its wings, almost all of the power is produced by the large muscle that is attached to the forearm at

Hummingbirds are named after the humming sound that is produced by their wings beating very fast. Some hummingbirds can produce over 90 wing beats per minute. Hummingbirds are very efficient at hovering but can only achieve this by being very small and very muscular.

one end and to the base of the breastbone at the other. However, birds do not just flap their wings when they fly, they also twist their wings and alter the angle of the feathers. On the down beat, the edges of the feathers overlap slightly so no air can get between them. This allows the wing to push against the air to get maximum lift. As the wing is brought up, the feathers are twisted slightly so that they separate, allowing air to move between them. This reduces air resistance. To hover, birds flap their wings forward and backward. Lift is produced by pushing air downward on both forward and backward beats of the wing. Over 50 wing beats per minute are needed for hovering, so many large birds are unable to hover because they cannot beat their wings fast enough.

Most birds are quite small compared with mammals and reptiles. There is a reason for this. If you could double the size of a sparrow, its weight would increase by eight times, but the strength of its flight muscles would only increase by four times. The amount of strength gained would not be enough to compensate for the increase in weight, and so the giant sparrow would not be able to fly so well. In general, the larger birds that fly have adopted gliding and soaring flight rather than flapping flight because gliding and soaring use up less energy. Two birds that are so heavy that they cannot fly, the ostrich and the emu, have very powerful legs that give them great speed on the ground.

The world's smallest hummingbird weighs about 2 g (.07 oz.) and has a wing span of less than 3 cm (1.2").

Gliding

Unpowered flight requires a lot of lift if the animal or machine is to remain airborne for any length of time. Gliding birds and human-made gliders have long wings to generate the maximum possible lift. Once in the air, they can use rising warm air currents called thermals. By circling within a thermal, they can gain extra height, and then glide on, losing height until the next thermal. Many of the heavier birds that have long wings, like vultures and storks, have difficulty getting into the air but have elegant soaring flight. The wandering albatross has wings 3.45 m (11.4 ft.) across and flapping them is almost impossible. By using air currents above open water, it can stay airborne for hours without moving its wings.

The albatross can sometimes glide for as long as six days without having to flap its wings once.

The albatross (above) and gliders (left) have a very similar shape that is ideal for gliding.

Special shapes for special jobs

Fighter aircraft can reach speeds in excess of 2,660 km/h (1,662 mph), about two and half times the speed of sound (Mach 2.5). Some research and surveillance aircraft can go even faster.

Look at the bird and plane silhouettes below. What type of flight do you think each is designed for? Choose from:
1 long flights at high speed
2 slow, gliding flight
3 low-speed take-off with long cruising flight
4 fast flight with high degree of maneuverability

Evolution has produced a wide range of wing designs for birds, each suiting a particular way of life. Birds differ from most aircraft in that they have the ability to change the total area and shape of their wings. Outstretched wings give lift. Wings folded back on the body reduce lift for swooping. Birds of prey such as hawks and falcons fold their wings back close to their body to increase the speed at which they dive. They can dive very fast from great heights to catch smaller birds.

The latest idea for high-speed aircraft is the mission-adaptive wing. Electronic controls will continually alter the shape of the wing to give the most effective shape for any speed or altitude. There will be no flaps or spoilers to interrupt the flow of air.

A

B

C

D

Controlling flight

EXPERIMENT

Fly a slow roll

It is possible to design a paper glider that will complete a slow roll, that means to fly horizontally while turning 360 degrees. You will need some stiff paper and glue.

1 Take a piece of stiff paper and fold it in the way shown below.

2 Stick the sides of the main fold together.

3 By folding the end of one wing up and the other down it will be possible to make the glider roll. You will have to experiment with the angle of the bend in order to succeed. Why do you think that bending the wings in this way should cause the glider to roll?

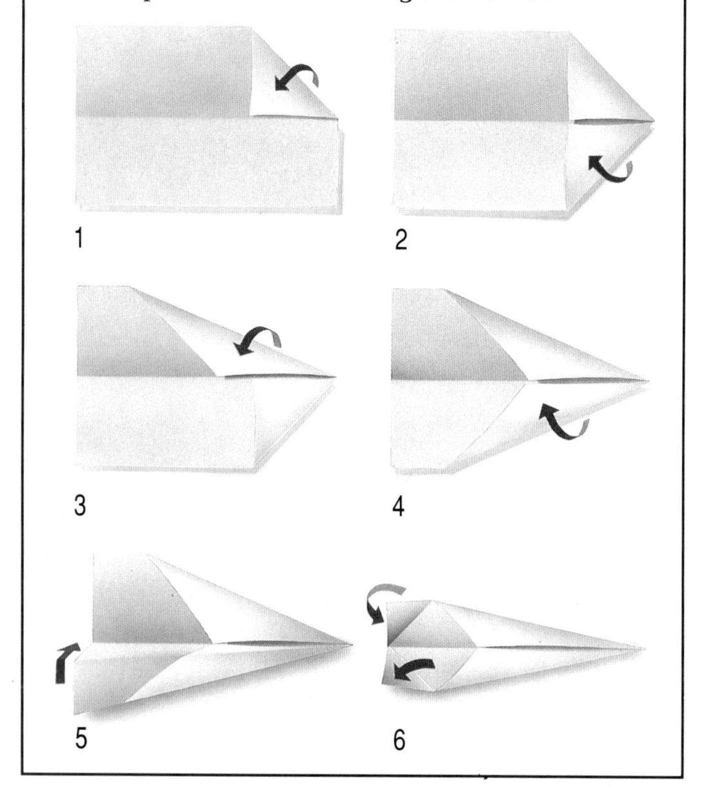

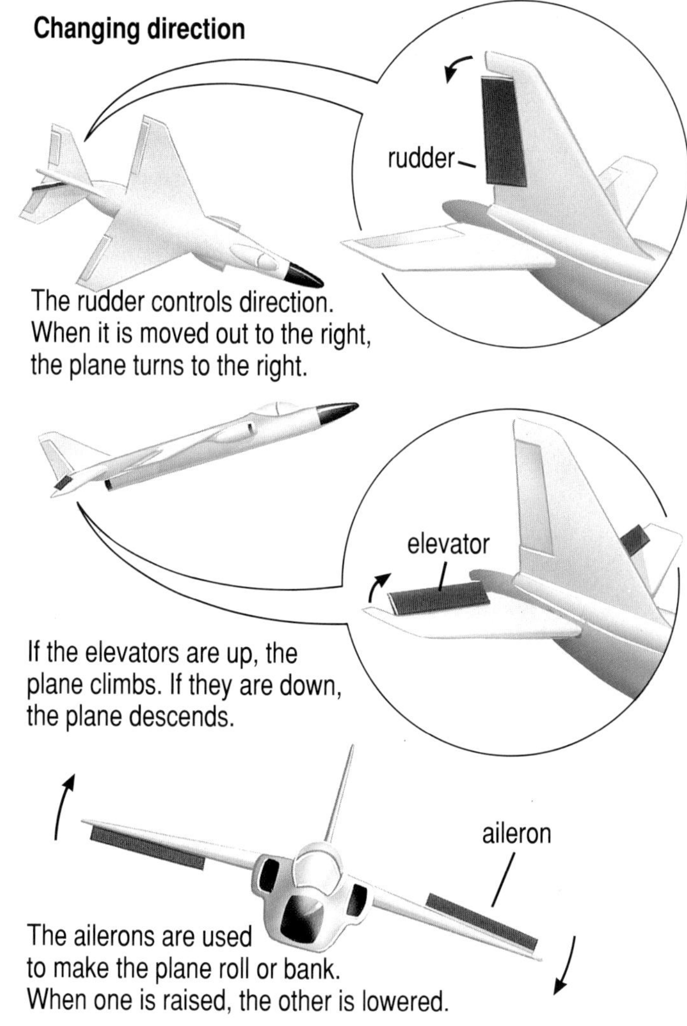

The rudder controls direction. When it is moved out to the right, the plane turns to the right.

If the elevators are up, the plane climbs. If they are down, the plane descends.

The ailerons are used to make the plane roll or bank. When one is raised, the other is lowered.

Once in the air, it is obviously very important to be able to change direction or move up and down. Birds use their muscles to adjust the shape and angle of wings and tail to control movement. Aircraft use ailerons and elevators to alter the shape and angle of their wings and so control movement in the air. To control the aircraft's direction, the rudder at the back of the tail can be angled from side to side.

Powered flight

Most aircraft are powered by engines. Small aircraft usually have propellers driven by piston engines, while modern passenger aircraft usually have jet engines.

Jet engines have a fan at the front of the engine that spins and draws air into a confined space. This compresses the air and so heats it up. The hot compressed air is passed into a combustion chamber where fuel vapor is sprayed into it. The fuel ignites and

A turbofan jet engine

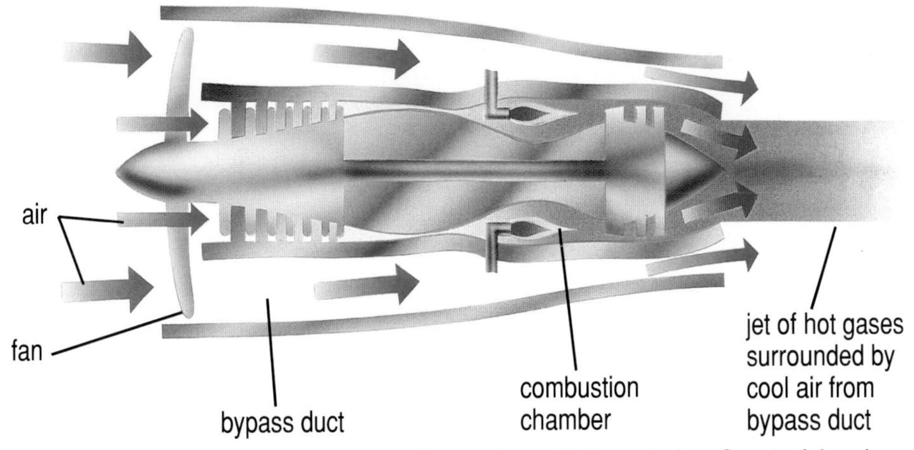

Many large modern planes use a type of jet engine called a turbofan. Some of the air drawn into the engine bypasses the combustion chamber. This air and the hot gases that leave the engine combine to provide greater thrust than in a conventional jet engine, which in turn improves fuel efficiency.

The squid uses a jet of water to produce movement.

expands, forcing hot gases out of the back of the engine. As hot gases leave the engine, they spin another set of fans, which drive the compressor fans at the front. It is the force of the hot gases leaving the engine that produces thrust, pushing the aircraft forward. You can see how escaping air produces thrust by blowing up a balloon and letting it go.

It is not only air that can produce thrust. Forcing water out through an opening produces the same effect.

The squid moves by drawing water in and then forcing it out of its body through a special opening. This jet of water pushes the squid in the opposite direction to that of the jet of water. By pointing the jet backward, the squid can dart forward to seize its prey. When the jet is directed forward the squid shoots backward, perhaps to escape when being attacked.

Take-off and landing

A mute swan taking off

Small birds can take off from a standing position on the ground, while larger ones must run to gain lift. Some launch themselves into the air from the top of a tall tree or cliff, dropping at first to trade off height for speed. As the speed of the air over the wing increases, more lift is obtained (see page 18). Taking off from water is more tricky. Water birds frequently use their large webbed feet to "run" across the surface of the water, while flapping their wings to gain enough lift to take off.

When landing, birds have to reduce speed. They spread their wings and move them into a more vertical position (near to stalling). They beat their wings against the direction of flight, to slow down, and stretch out their legs. Water birds brake by skidding to a halt on their webbed feet. Landing is quite a risky business, especially for some of the largest and strongest fliers, which have poorly developed legs because they spend so much time in the air. The blue-footed booby, a large sea bird, often crashes when it comes in to land.

Are there any similarities between the Concorde and birds in the way they land? What about other types of aircraft?

When a large aircraft takes off it needs maximum thrust and lift to gain height as quickly as possible. The engines are at full power. Aircraft wings have flaps next to the ailerons. These can be moved out to make the wing larger to give more lift during take-off, and pulled back once the aircraft has gained height.

When landing, the aircraft has to slow down as much as possible while still maintaining lift. The flaps are angled upward to create drag and reduce speed. Spoilers on the wings are raised during landing to further increase drag.

Can you think of any similarities between the Harrier Jump Jet and the squid?

Wings designed for high-speed cruising do not give very much lift at low speed, so many modern aircraft need long runways to build up sufficient take-off speed. However, not all aircraft need a long runway for take-off and landing. The De Havilland Dash 7 has wings that give maximum lift at low speed. This type of aircraft is called a Short Take-Off and Landing aircraft (STOL). Vertical Take-Off and Landing aircraft (VTOL), such as the Harrier, were designed for use in areas where there is restricted space, such as on aircraft carriers and in battle zones. The lift needed for take-off is produced by four nozzles that direct hot gases from the engines downward.

The structure of wings

The frigate bird has a wingspan of over 2 m (6.6 ft.), but its skeleton weighs only 125 g (4.38 oz.).

Airborne structures must be both very light and very strong to be efficient. The skeleton of a bird and the airframe of an aircraft are similar. They are both light, rigid and strong, in order to take the stresses of flight.

Birds are light in weight. Their bones are hollow, with internal struts for added strength. Their lungs, a series of air-filled sacs, take up more body space than in a mammal of the same size.

The design of the honeycomb combines great strength with minimum weight.

Aircraft wings are hollow, just like a bird's bones. Rows of ribs are linked by spars and covered with an outer skin of thin metal. The spars are pierced by many holes. Aircraft floors and other components are made from a honeycomb mesh with a thin sheet of metal glued to each face, just like a natural bee honeycomb. The honeycomb shape is a highly efficient design — it is very strong but uses the minimum amount of material.

The feathers of a bird's wing give a very smooth surface and enhance laminar flow over the wing. NASA has attempted to improve the speed and efficiency of supersonic aircraft by improving the laminar flow over the wings of a jet fighter. The wings are fitted with an experimental skin that has millions of tiny holes drilled in it. A suction pump draws off turbulent air above the wing down through the holes, leaving a smooth flow of air over the wing. If trials are successful the idea could be used on high-speed aircraft in the future.

Parachutes and helicopters

A human-made parachute has a large canopy. This produces a huge surface area, up to 50m^2 (60 sq. yd.),which creates a lot of air resistance. The canopy is attached by rigging lines and a harness to the parachutist.

A parachute is a structure designed to slow down the rate of descent. Without a parachute, a person will free-fall at 50 m (165 ft.) per second; with a parachute this can be reduced to only 5 m (16.5 ft.) per second. As the parachutist falls through the sky, the air resistance increases. Eventually this force is balanced by the weight of the parachutist. A parachute needs to have an air hole at the top of the canopy to allow the air to pass through. If there is no hole the parachute will swing from side to side as the air escapes from the edges.

Many plants use "parachutes" to disperse their seeds. The seeds are released when there is a light wind to carry them away. Feathery hairs project from each seed to increase its surface area. This increases the air resistance so that the seed sinks slowly to the ground, which makes it more likely to spread further away from the parent plant.

EXPERIMENT

Make your own parachute

You will need a square of cotton material, about 25 cm x 25 cm (10" x 10"), some cotton thread and a small model figure (or peg or some Plasticine).

1 Cut four equal lengths of cotton thread, each about 50 cm (20") long.
2 Tie one length of thread to each corner, and then tie all the other ends together.
3 Finally, tie the model figure to the knot you have just made.
4 Fold the parachute, with the figure inside it, being careful not to tangle the thread.
5 Throw the parachute up into the air, and watch it open and fall to the ground.

Does the parachute fall straight down, or swing from side to side?
How could you improve the design of the parachute?
How could you improve this experiment?

Why is it important for seeds to be dispersed away from the parent plant?

Pollen being released from the catkins of the alder tree.

Many flowers rely on wind pollination. This means that they use the wind to carry the pollen from the male flower to the female flower so that fertilization takes place and seeds can form. The pollen grains of the pine tree have air sacs, making them very light. The pollen is carried in the air for huge distances. Air samples taken from 5–6 km (3.125–3.75 mi.) above the earth contain pollen from flowers and spores from fungi.

Helicopters have rotor blades that provide enough lift to support the weight of the machine. Each wing-shaped blade generates lift as it drives through the air. The tail rotor prevents the body from spinning in the opposite direction to the main rotor.

Many seeds fall in helicopter style. They spin as they fall. This slows down the rate of descent greatly so there is more time for any breeze to blow the seeds away from the parent plant. The maple has seeds like this. The seeds are formed in pairs. The membrane covering them extends either side to form a pair of wings. If the seeds break apart, they still spin.

These maple seeds will soon drop from the tree.

Helicopters get lift from rotating blades above the cabin.

EXPERIMENT

Helicopter seeds

You will need some maple seeds, a piece of stiff card, some Plasticine and a stopwatch.

1 Take some of the maple seeds.

2 Record the time it takes for a seed to drop a certain distance, for example 2 m or 6 ft.

What happens if you break one of the double seeds in half and try again?

3 Using the card and Plasticine, try to make your own helicopter seed shape.

Is there a relationship between the weight of the seed and the length of the wing? Can you find any other seeds that behave like helicopters?

Insects and bats

Are there any similarities in the way lift is produced by insects and the Harrier aircraft? (See page 24.)

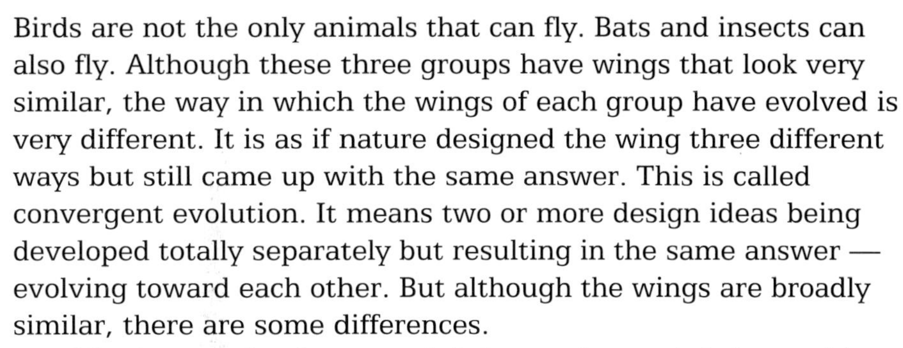

Birds are not the only animals that can fly. Bats and insects can also fly. Although these three groups have wings that look very similar, the way in which the wings of each group have evolved is very different. It is as if nature designed the wing three different ways but still came up with the same answer. This is called convergent evolution. It means two or more design ideas being developed totally separately but resulting in the same answer — evolving toward each other. But although the wings are broadly similar, there are some differences.

The insect wing is unusual. It has no bones; it is just a thin membrane supported by hollow veins. When a cross-section of an insect wing is examined closely, it is found that it resembles folded paper. How can this wing section possibly create lift? It does not look anything like an airfoil. By doing tests in a wind tunnel, scientists have discovered that, in fact, the insect wing does behave like an airfoil after all. Each fold causes a small eddy, or turbulent flow area, to form in the depth of the fold. This causes the air to flow smoothly over the top of the eddies, just like in an airfoil. This is very similar to the laminar flow over a dolphin (see page 13).

The dragonfly has two pairs of wings. When the first pair of wings is on a down beat, the second pair is on an up beat.

The dragonfly is a large insect and a powerful flier. This insect can glide but smaller insects cannot. Small insects have to flap their wings continuously in a figure eight. As the wings twist, they produce thrust on both the up and down strokes. Some insects beat their wings up to 1,000 times each second, producing a stream of air down and back. To climb, an insect inclines its body.

The bat wing is also very different from that of the bird. It is formed by a thin layer of skin stretched across long, thin arm and finger bones and down to the hind leg. The thumb remains as a claw on the top of the wing. The bat has a deep chest like the bird and its bones are thin and very light. By altering the tension in its skin the bat is able to control the shape of the wing and hence its speed. Bats can maneuver well at low speeds.

The largest ever flying insect was a dragonfly that lived over 300 million years ago. Its wing span was over 50 cm (20").

The closest that humans have come to copying the wing of

The common midge can beat its wings at over 57,000 times per minute, reaching 133,000 beats per minute in short bursts.

Bats and hang gliders both have wings that consist of a thin membrane stretched across a tight frame.

the bat or bird is in the wings of hang gliders and microlight aircraft. In these ultra-light craft, the wings are formed from a thin sheet of nylon stretched across a frame of hollow aluminium tubing and are braced by wires.

Lighter than air

How do you think a blimp or zeppelin can increase and decrease its height in the air?

Some gases are lighter than air. Hot air balloons are made of a huge envelope of lightweight material such as nylon attached to a basket called a gondola. There is a gas burner hung under the envelope to heat the air in the balloon. The heat causes the air to expand and become less dense, and so become lighter than the surrounding air. If the balloon, its air and everything it carries weigh less than the displaced air, then the balloon will rise. Every cubic meter of air weighs 1.3 kg (2.86 lbs.), so the gas in a balloon about 20 m (66 ft.) in diameter displaces about 408 kg (897.6 lbs.) of air. If the gas in the balloon is half as dense as air, then the balloon and its occupants can weigh up to 200 kg (440 lbs.) and the balloon will still rise.

Helium is almost eight times lighter than air, so it is very good at producing lift, but it must be sealed in to prevent it escaping. As the balloon rises, and the outside atmospheric pressure decreases, the gas in the balloon can expand, so the balloon will eventually burst unless the gas is gradually released.

Richard Branson and Per Lindstrand were the first people to cross the Atlantic Ocean by hot air balloon, in July 1987. It was the world's largest balloon, with a volume of 65,000 cm³ (22,750 cu. ft.).

Balloon pilots have no control over the direction a hot air balloon travels; they go wherever the winds take them.

Airships (blimps and zeppelins) are generally much larger than hot air balloons. They consist of a huge cigar-shaped envelope filled with gas. Early designs used hydrogen, which was flammable in oxygen and caused several spectacular disasters. Helium, however, does not burn in oxygen and so is quite safe. Propellers at the back provide thrust, and by varying the power to the propellers, the pilot can steer the airship. The heaviest parts of an airship are the engines. Everything else is made of very lightweight materials. The cables attaching the gondola to the envelope are made from Kevlar, which is a new kind of plastic. It is stronger than most steels but is only one fifth the weight. Because they do not require much power to stay up in the air, airships are particularly useful in jobs where they have to stay in the air for a long time, such as surveillance.

Hot air balloons (below) and blimps (below right) both rely on gas to provide lift.

Key words
Airfoil wing shape that cuts through air and produces lift.

The Hindenburg, *a zeppelin, built in 1936, weighed 236 tonnes (260 tons), and was filled with 200,000 m³ (7,000,000 cu. ft.) of hydrogen. It could travel for 20,000 km (12,500 mi.) without refueling. It was totally destroyed by fire in 1937.*

Moving on land

Many machines use wheels for movement on land, but wheels cannot move over all surfaces.

Nearly all terrestrial animals have solved the problem of moving on land by evolving jointed limbs. Muscles attached to the skeleton move the limbs, which act as levers and push backward against the land to produce forward movement.

There are many similarities between the designs of animals and machines that move in air and water. On land, most human-made machines use the wheel, but no animal uses wheels for locomotion. Why should this be so?

A wheel works by means of an axle rotating freely in a bearing. The key difference between this and animal design is that most animal joints can only rotate through a limited angle because of the need to connect the limb to the rest of the body. This makes it extremely unlikely that animals could evolve a structure like an axle, which relies on not being connected to the bearing.

The squirrel is a particularly agile animal. It can run, climb and jump.

This robot insect mimics the design of a real insect. Its jointed legs allow it to move over bumpy surfaces.

While wheels are particularly well suited to moving on flat surfaces, jointed limbs like arms and legs are much more versatile. They allow animals to climb trees and even mountains, which a wheeled vehicle cannot do. This is another good reason why animals have evolved in this way. Some experimental robots designed by scientists to move over very rough terrain have been equipped with legs that work like a spider's legs.

Even though limbs and wheels are quite different solutions to the problem of movement on land, the forces that affect them are the same. The most important of these are gravity and friction, and for those objects that move fast enough, air resistance can be a factor. It is an important consideration when designing high-speed vehicles such as cars and trains.

Gravity

Gravity is a force between two objects that attracts them to one another. The force depends on the mass of the objects, and how far apart they are. The larger the mass of an object, the stronger is its gravitational pull. Objects that are close together have a greater attraction than the same two objects far apart. Gravity is a weak force, and it requires a very large mass before its effects can be detected. Although there is a gravitational pull between two people standing side by side, the attraction is so tiny that it cannot be measured. However, the planet Earth has a huge mass and so it has a large gravitational pull. It is strong enough to hold us firmly

on the ground. If we drop something, it falls to the ground because it, too, is attracted toward the Earth, pulled by the force of gravity.

The moon stays in orbit around the Earth, and the Earth around the sun, because of gravity. When objects fall due to the Earth's gravity, their speed increases at the same rate, up to a maximum of 9.81 m per second per second (9.81 m / s^2). Scientists refer to this number as g.

All things, when dropped, fall to the ground, pulled by gravity.

The gravitational pull of the moon is only one sixth that of Earth. If you walked on the moon your mass would be the same but your weight would be one sixth of your weight on Earth.

Friction

?

Why do gymnasts rub powder on their hands before they work on bars or rings?

When two surfaces rub together, a frictional force is produced that resists the motion. You can feel this force for yourself by pushing an object such as an eraser across a table. If you press harder, the friction increases, and the object becomes harder to move. Even if the surfaces of the table and the object are very highly polished, there is still some friction. Surfaces might appear to be completely smooth, but on a microscopic scale most are still very rough. In general, the rougher the two surfaces, the greater the friction.

Friction is produced as the ridges on the two surfaces rub against one another, and a force is required to move one over the other. Any energy put into overcoming friction is lost as heat. That is why your hands warm up when you rub them together. Since friction occurs in all machines, no machine is 100 percent efficient, as some energy is inevitably lost as heat.

Although friction can cause problems, it is essential to the way we live. For example, friction between the tires of a car and the road surface is necessary so that the wheels do not spin. When brakes are applied to the wheels, friction is produced that slows the car down. Friction helps to prevent our feet from slipping when we walk, and helps to hold things in place when we put them down. Thousands of everyday

A magnified view of the rough surface of a piece of sticky paper

Runners rely on friction to stop them from slipping.

activities would simply be impossible without friction.

Considering the importance of friction, it is perhaps surprising that so many of our efforts go into reducing it. For example, we try to minimize friction in machinery in order to reduce both power consumption and wear on the moving parts.

Friction can be reduced in a number of ways. The first is to make the bearing surfaces from a low-friction material such as brass, bronze or nylon. Another way is to separate the two surfaces with a lubricant such as oil or grease, the molecules of which slip easily over one another. Finally, an intermediate object can be used to roll between the two surfaces. The rolling object is known as a bearing. Ball bearings are smooth round objects that roll over each other but do not rub, so keep friction to a minimum. They can be lubricated with oil or grease to make them roll even more smoothly.

Why does a car need a longer braking distance in wet weather?

EXPERIMENT

Comparing friction on different surfaces

You will need two pieces of plywood, some strong sticky tape, a selection of different materials to cover the wood — such as felt, a thin sheet of metal and a sheet of plastic — a small toy car or small block of wood, some Fun-tak (a puttylike substance used to put up posters) and a protractor.

1 Hinge the two pieces of wood together using strong tape.
2 Cover the slope with the first test piece of material.
3 Stand the protractor beside the hinge and secure it in place using a small amount of Fun-tak. Lift one piece of wood so the angle of the slope is 10°.
4 Place the car or block at the top of the slope. Does the car or block move?
5 Slowly increase the angle of the slope by 10° at a time. Record the angle at which the car begins to move.
Try this with other materials. How can you tell which material produces the most friction?

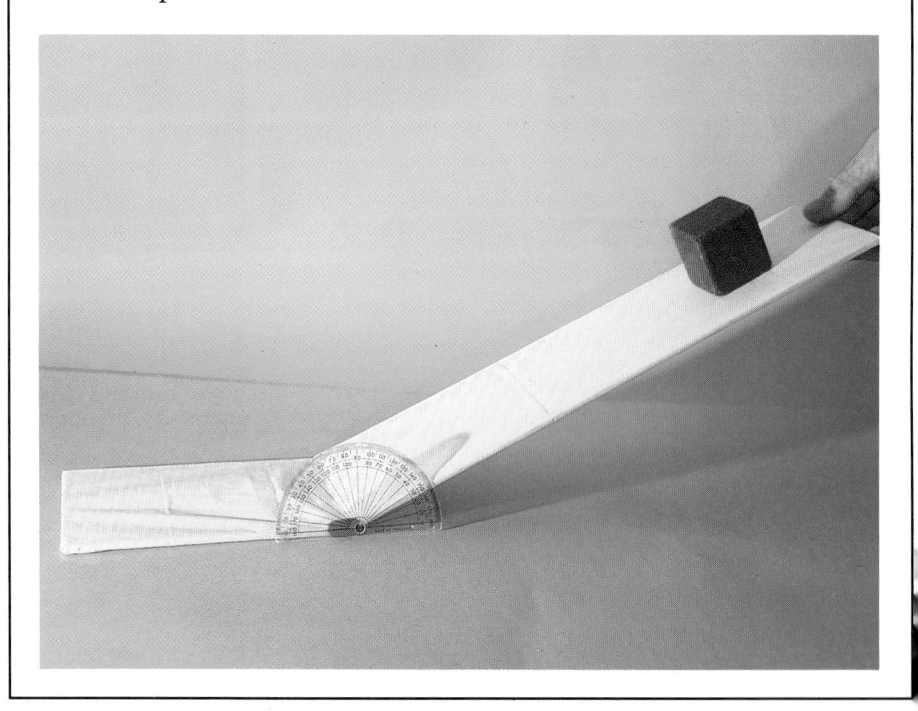

Air resistance

The cheetah is the fastest land animal. How is it adapted to running at high speeds?

An object moving fast requires work to be done in order to overcome the frictional drag of air. Air resistance can be quite significant for humans when they run. The cost of overcoming air resistance for running a middle distance race is 7.5 percent of the total energy cost. A sprinter may use over 13 percent because air resistance increases with speed. Today, sprinters wear figure-hugging body suits made from an elastic fiber. The suits produce

The fastest 100 m sprint on record is 9.92 seconds.

less air resistance than T-shirts and shorts, and also help to keep leg muscles warm, so lessening the chance of injury.

Running behind another runner also reduces air resistance, which is why tactics are so important in middle distance races, such as the 800 m and 1500 m (sprints are too short for such tactics to become important). However, walking incurs little air resistance. So for most animals, which do not usually move at high speeds, air resistance costs are minimal and little adaptation is needed to reduce drag.

Car bodies are designed so that air flow is as smooth as possible. For a car traveling fast, air resistance is by far the largest force trying to slow it down. Friction between the air molecules as they are forced aside and move over the car, and turbulence in the airflow behind it, both contribute to drag. Car manufacturers often quote the coefficient of drag, or Cd, of a car. This is a measure of the car's resistance to moving through air. For example, a car with a Cd of 0.37 is more streamlined than one with a Cd of 0.55. Streamlining reduces the air resistance, improves the performance and economizes on fuel. Grand Prix racing cars have a low, smooth streamlined shape to reduce air resistance and a rear spoiler to produce a downforce to improve roadholding at high speed.

This car has a very streamlined shape.

Elastic forces

Some things change shape when they are pushed or pulled. We say that they deform. If the object returns to its original shape when the force is removed, we say that it is elastic. A rubber band is elastic. So is a spring.

The springs of a car support its weight but allow the wheels to move rapidly over bumps in the road. The springs absorb energy as they are compressed, and spring back into shape once over the bumps. This helps the body of the car to travel more smoothly, providing a better ride for the passengers.

When a rubber band is stretched, it stores energy. When it recoils back into shape again, most of this energy is converted back into movement energy, and the rest is lost as heat.

?

Can you think of other ways in which elasticity is used?

The muscles and tendons in the bodies of animals are also elastic. This elasticity helps animals to move. Kangaroos cannot walk in the same way as other animals, but they use elasticity to

The storage and return of energy is so efficient for the kangaroo that it uses the same amount of oxygen at top speed as it does at half speed.

help them jump. When they are moving slowly, they use all four feet and the tail. Their short forelimbs help support the body as the rear limbs hop. When moving faster, they use only their hind legs, which have very long Achilles tendons that act like gigantic rubber bands, storing and returning energy on every hop. As they reach top speed, the rate of jumping does not increase, only the length of each jump.

Elastic recoil helps to absorb shocks. Why might this be important?

!

The Klipspringer, a small South African antelope only 50 cm (20") tall, can jump 10 m (33 ft.) up nearly vertical cliffs.

EXPERIMENT

Knee bends

Elastic recoil is important for movement. You can discover this for yourself by carrying out this simple exercise.

1 Stand up with your feet together.
2 Slowly bend your knees, keeping your back straight, until you are squatting with your hands touching the floor beside you.
3 Hold this position for a few seconds, then stand up.
4 Repeat steps 1 to 3 nine more times with a short pause at the bottom of each bend.
5 Take a brief rest.
6 Now you should repeat steps 1 to 4 again, but with no pause at the bottom of each bend.

Was it any easier with or without the pauses?

Explanation By not pausing at the bottom of each bend, you can take advantage of elastic recoil to bounce up a little, which saves some of the energy needed to return to the standing position. If you pause at the bottom, the elastic energy is lost.

The tree frog uses powerful leg muscles to thrust it foward as it leaps.

Fleas and grasshoppers can jump over 50 times their own body length. If we could jump as well as fleas can, we could leap a 30-story building. So why can't we jump as high? It is all a question of scale.

The average grasshopper weighs 1 g and can jump 50 cm. If we were to breed a giant grasshopper, say 10 times longer, wider and taller, then its mass would increase by 1,000 (10 x 10 x 10) times to one kilogram. However, the cross-section of the area of its muscles would only increase by 100 times (10 x 10). Since it is the muscle cross-section that determines the insect's power, there would be proportionately less power to move the mass — one tenth as much, in fact. So our poor old giant grasshopper would only be able to jump 5 cm.

The very long back legs of the grasshopper are ideal for leaping.

Fleas use the principle of the catapult to jump. They store energy in an elastic material at the base of their hind legs. It is called resilin and has the same properties as rubber. The flea's muscles compress the resilin and then release it, providing a slingshot effect that acts much faster than muscles could contract or expand. The flea can accelerate at up to 2,000 m/s^2 — over 200 times the acceleration due to gravity, g (see page 31). A human being would pass out at an acceleration of 10 g.

The click beetle uses a similar mechanism to turn itself right-side-up if it lands on its back. After struggling to right itself for a few moments, it stops moving, and springs into the air with an audible click. If it lands upside down again, it simply repeats the process until it is righted. In the jump, it reaches heights of 30 cm (12") or more. How does it do this? On the main part of its body there is a peg pointing backward that fits into a pit or recess. When the beetle is resting, the peg is held in place under tension, but when released, the peg flies out like a little hammer, providing the force needed to propel the beetle up into the air.

The latest running shoes have air-cushioned soles. The middle of the shoe has a small pocket of gas that provides much more cushioning than foam alone.

Elastic recoil is used in the design of running shoes. During a marathon, a runner's feet hit the ground over 25,000 times, each impact equivalent to a force of two and a half times the body weight. The repeated pounding of the runner's foot on the road can lead to severe injuries. Running shoes are designed to reduce injuries by use of elastic material to lessen the impact and return some of the energy (just like elastic recoil).

Levers and joints

If you had to loosen a nut, would you find it easier with a long or short wrench?

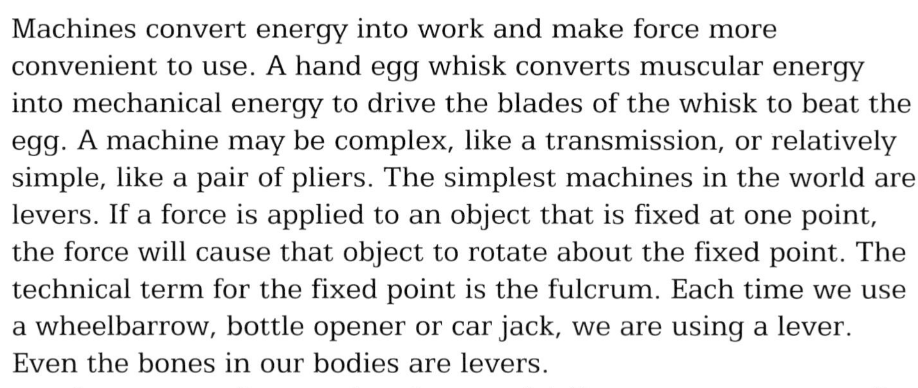

Machines convert energy into work and make force more convenient to use. A hand egg whisk converts muscular energy into mechanical energy to drive the blades of the whisk to beat the egg. A machine may be complex, like a transmission, or relatively simple, like a pair of pliers. The simplest machines in the world are levers. If a force is applied to an object that is fixed at one point, the force will cause that object to rotate about the fixed point. The technical term for the fixed point is the fulcrum. Each time we use a wheelbarrow, bottle opener or car jack, we are using a lever. Even the bones in our bodies are levers.

Levers are often used as force multipliers, to convert a small force into a greater one. For example, a heavy load can be moved a short distance by applying a small force to a crowbar. The longer the crowbar, the easier it is to move the load.

The rudder of this barge is easily controlled using a lever.

There are three classes of lever, depending on the relative positions of the fulcrum, load and effort. Class one levers have the fulcrum in between the load and the effort. They are the most efficient form of lever. Typical class one levers are scissors and crowbars. Class two levers have the fulcrum at one end, and the load acts downward between the effort and the fulcrum. Nutcrackers and wheelbarrows are class two levers. Class three levers have the fulcrum at one end and the load at the other, with the effort in between them. They are the least efficient class of lever. The human forearm is a good example of a class three lever.

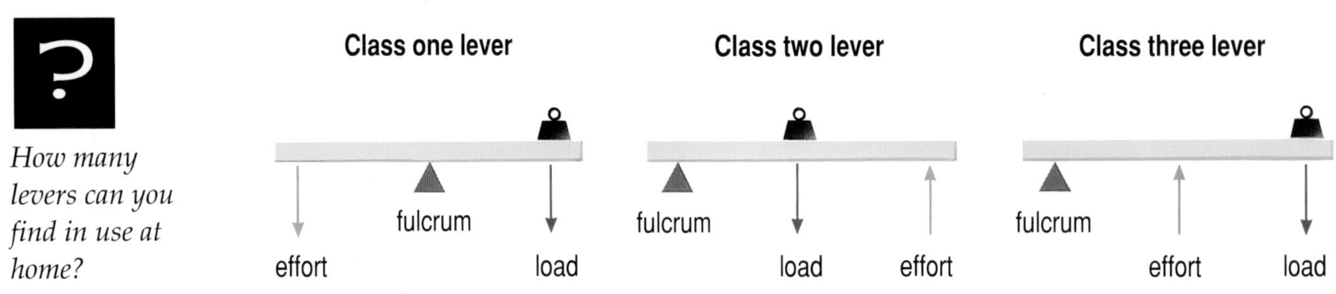

How many levers can you find in use at home?

EXPERIMENT

The seesaw

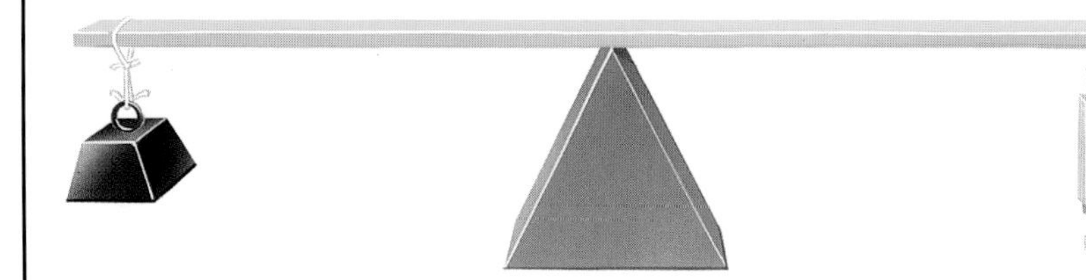

You will need a long ruler, different weights, a small spring balance and a triangular piece of wood.

1 Set up the seesaw as shown in the diagram. Attach a weight at one end and a spring balance at the other.

2 Use the spring balance to measure how much force is required to lift the weight.

What happens if you alter the position of the balance? Is it easier or more difficult to lift the weight if the balance is moved nearer the fulcrum?

Joints in animals

Can you think of any examples of human-made ball-and-socket joints?

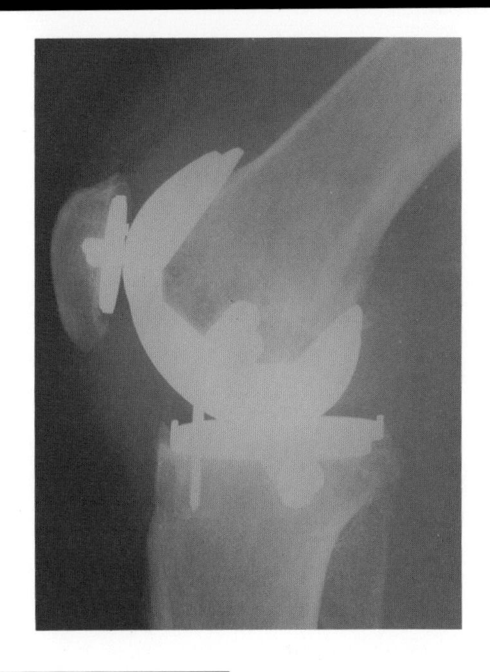

Nowadays it is possible to repair worn joints with new, human-made joints.

In bony vertebrates a joint is formed where two bones meet. Joints are essential to give the skeleton flexibility, for without them the skeleton would be immobile. The skeletons of mammals have three main types of joint. Fully movable joints, such as the ball-and-socket joints of the shoulder and hip, and the hinge joints of the knee and elbow, are the most familiar. These joints are the most important to movement. They are designed so that there is minimal friction between the two bones. The ends of the bones are covered by a very smooth material called cartilage. A lubricant called synovial fluid separates the two surfaces. Partly movable joints, the third type of joints, are found in, for example, the backbone, where less flexibility is required. Finally, a few joints are immovable, such as the joints between the skull bones, which are fused together.

Joints suffer wear and tear, just like moving parts in machinery. Many elderly people are affected by a condition called arthritis, when joints become swollen and painful and flexibility is reduced. Forty-nine out of every 50 people suffer from arthritis at some stage of their lives. There are many kinds of arthritis but the most common is osteo-arthritis, in which the cartilage becomes worn and split and the underlying bone becomes thicker. The joint swells and stiffens, until eventually the ends of the bone are exposed and rub on each other, causing severe pain. This is very common in finger joints and the weight-bearing joints of the hip and knees. Nowadays it is possible to replace severely damaged joints with artificial ones made from plastic or metal.

Key words

Friction the force that resists movement when one surface slides over another, producing heat energy.

Gravity the force of attraction between matter. It is the force that holds us on Earth.

Elasticity the tendency of some materials to return to their original shape after being stretched or squashed.

Fluids and pumps

Hydraulics at work

Hydraulic systems are used in heavy machinery. The arm of this digger is controlled by pushing on the fluid in the tubes.

Hydraulics — the movement of fluids in pipes — is based on very simple principles. The most important principle is that liquids cannot be compressed. When a force is applied to a liquid the force is transmitted through the liquid with no loss. In fact, the pressure is transmitted equally to all parts of the liquid. If two pistons of the same size are connected by a pipe, and filled with a liquid, then when the "input" piston is pushed in by a certain amount, the "output" piston will move out by the same amount.

Hydraulics allows us to exert a force at one point that has its effect at a distance. Although this is useful in itself, a more useful feature of hydraulic systems is the ability to magnify force. If the output piston has a larger area than the input piston, then it will not move as far as the input piston but the force exerted at the output end will be greater. It is really very similar to using a lever.

EXPERIMENT

An artificial braking system

You will need four small syringes and one large syringe, some rubber tubing, and a few T pieces to join the tubing. (Use syringes *without* needles, made for use in ears.)

1 Connect the five syringes as shown in the diagram.

2 Fill the system with water through the large syringe, keeping the plungers of the four small syringes fully in.

3 When you are ready, push the plunger into the large syringe.

What happens to the plungers of the four small syringes? Why has this happened? By how much did you push in the large plunger and by how much did the small plungers extend?

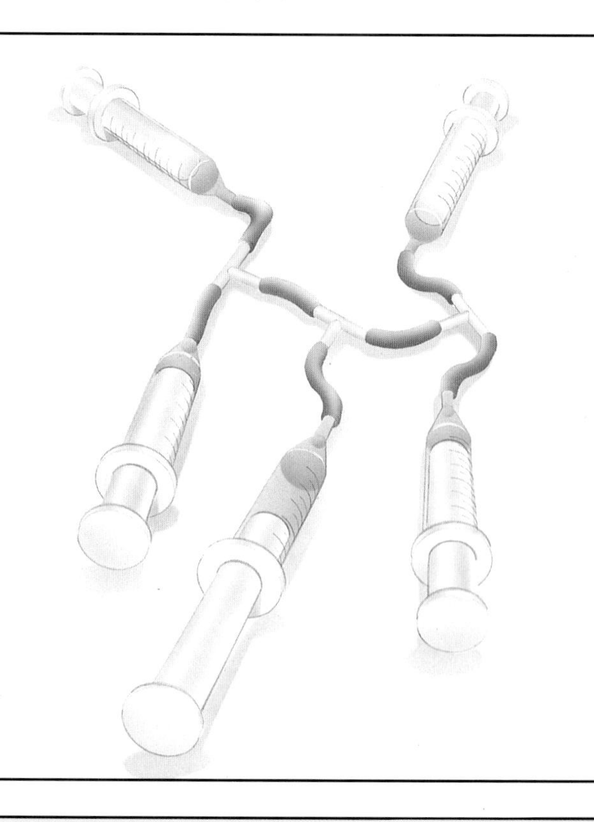

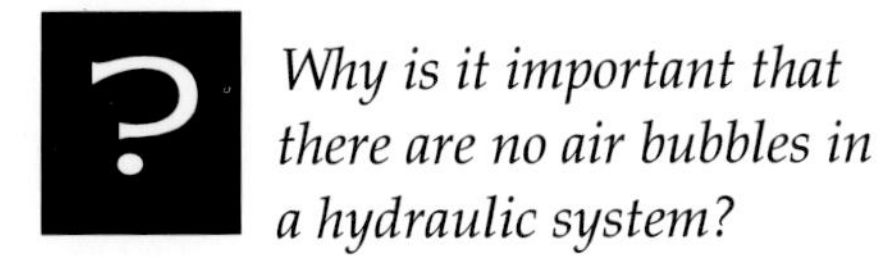

Why is it important that there are no air bubbles in a hydraulic system?

Hydraulic systems are used in many machines. The braking system of a modern car consists of one small "master" piston, pressed in by the brake pedal, connected to four much larger "slave" pistons, one at each tire. When the driver presses the brake pedal, pressure is applied to the hydraulic fluid, which in turn pushes on the slave pistons, each of which brakes one tire.

Hydraulics is particularly important in heavy industrial work such as building tunnels and roads, because it can be immensely powerful and yet is easy to control.

Hydraulics in the animal kingdom

Hydraulic systems are found in many animals. Spiders depend on hydraulics. They do not have any muscles to extend their legs. The only way they can do this is by applying pressure to their body fluids. They use their muscles to press on the fluid to push out their legs. If a spider is wounded or dies, it is unable to maintain enough pressure to extend its legs, so they curl up.

Starfish also use hydraulics. Along the underside of each arm are rows of tube feet. The tube feet are filled with fluid and can be extended by increasing the

The starfish has hundreds of tiny tube feet that can be extended by squeezing fluid into them.

The earthworm moves by pushing on fluid inside its body.

pressure inside. They are retracted by muscles. The base of each tube foot is like a small sucker, which grips on to rocks so the starfish can pull itself along. They feed mainly on bivalves such as mussels. The arms of the starfish wrap around the bivalve and the tube feet grip the shell.

Have you ever watched an earthworm move? This is hydraulics in action, too. The worm extends the front part of its body, and then pulls the back part up after it. Earthworms have a body wall with two muscle layers; one is circular and the other is longitudinal (runs lengthwise down the body). These muscles surround the body fluid. When the circular muscles contract, they increase the pressure on the body fluid, which in turn makes the body long and thin. Then the longitudinal muscles contract, pulling the back part of the worm forward. The worm gets shorter and fatter, moving forward as it does so.

Pumps

Screw pumps are operated by a handle that turns the screw. As the screw turns, it moves water from the bottom of the screw to the top.

Pumps are used to move fluids from one place to another. The heart is a pump, moving blood around the body. The water that flows out of a tap has been pumped from a water main.

All pumps require an external force, such as a motor or a muscle, to operate them. A pump can boost the power of the flow or lift the fluid to a greater height, or do both.

Simple hand pumps were in use as early as 3000 BC. They were simple lift pumps, used by farmers to raise water from underground wells into irrigation channels so that they could water their crops.

In a lift pump, a plunger is raised and water is drawn up behind it into a cylinder. As the plunger goes down, so the pressure of the water closes a valve at the bottom of the cylinder, preventing the water from escaping. Meanwhile, a valve in the plunger opens, letting water into the space above it. As the plunger is raised, the valve closes and water flows out.

A lift pump

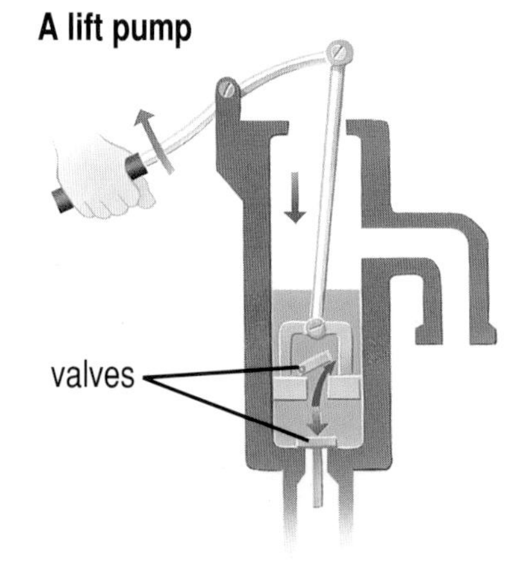

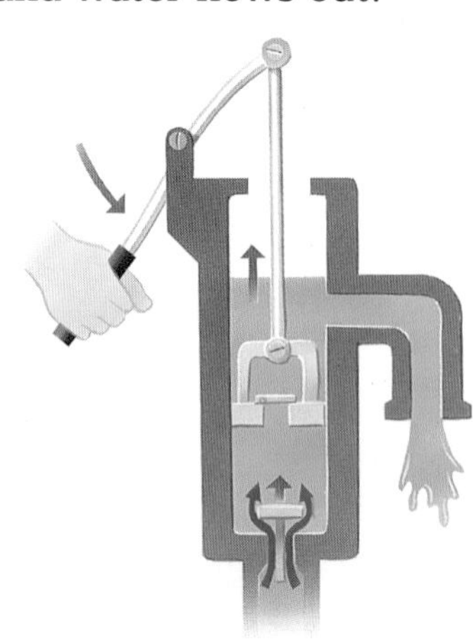

As the plunger is pushed down, the pressure of water closes a valve at the bottom of the pump and opens a valve at the top. This lets water in above the plunger. When the plunger is pulled up, the top valve closes so water is pushed out of the spout. At the same time, the bottom valve opens because the pressure of water is less, so more water is drawn up.

The heart, a natural pump

! *In an average human lifetime, a heart will beat over 2000 million times.*

Many of the more complex animals have a heart. Its job is to move blood around the body. Insects and worms have quite simple hearts but the most advanced ones are found in mammals. The human heart is actually two pumps, side by side.

But why do we need a heart in the first place? Small single-celled animals do not need a heart to move materials around. They have a large surface area across which oxygen and food materials can enter the cell directly. Our bodies are much larger and are composed of millions of cells. Oxygen and food have to be moved around the body by a fluid called blood. It is the pumping action of the heart that keeps the blood moving in the blood vessels. The heart has to be able to respond to changes in the body's work rate: if we start to exercise, we need more oxygen, so our heart has to work harder to pump blood around the body.

The body of an adult has 100,000 km (62,500 mi.) of blood vessels.

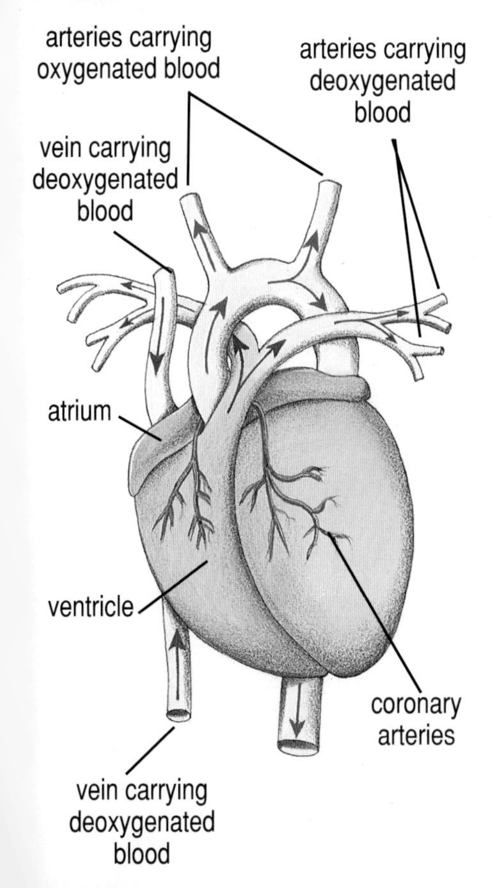

EXPERIMENT

Taking your pulse

All you need for this experiment is a stopwatch, or a wristwatch with a second hand.

1 Sit down and rest for a few minutes, then take your pulse. Feel for your pulse on your wrist or on the side of the neck just under your jaw. Count how many beats you can feel in 20 seconds.

2 Now get up and run around for two minutes, and immediately take your pulse again.

What has happened to your pulse rate? How long does it take to get back to what it was before?

The heart does not have any gears or levers, so how does it work? The heart is made mainly of muscle, called cardiac muscle, which beats throughout your life. The cardiac muscle provides the power for the heart. There are four spaces, or chambers, in the heart, two on each side. Each side is divided into an upper chamber, the atrium, and a lower chamber, the ventricle.

Blood returning from the body and lungs enters the atria. When the atria are full of blood, the muscles in the wall of the atria contract, squeezing the blood through a valve into the ventricles. Then the ventricle muscles contract, forcing the blood out of the heart into large arteries.

Our heart beats about 65-70 times every minute. Other mammals have different heart rates. Large mammals have a much slower heart rate than small mammals. The heart of a mouse will beat about 200 times every minute. Some scientists think that there may be a maximum number of beats which a heart can ever produce in its lifetime, and that this is one reason why small mammals have shorter lives than larger ones. They have the same number of heartbeats, but use them up more quickly.

? *In the human heart, why must the atria contract before the ventricles? What would happen if the heart valves were faulty?*

EXPERIMENT

Looking at a heart

You will need to buy a sheep or pig heart. Most butchers sell them. You will also need a chopping board and a pair of scissors.

1 Have a good look at the outside of the heart. Can you see any blood vessels? How large are they? You may be able to see the veins going into the heart and the arteries leaving the heart. The arteries are the big blood vessels at the top. They look like rubber tubes.

2 Using a pair of scissors, carefully cut through the wall of the heart and have a look inside. You should be able to see the large ventricles and the strings that are attached to the valves. The atria are very small and may not be present if the butcher cut across the top.

There is no way that blood from the right side of the heart can get through to the left. Can you remember how blood gets from one side to the other?

Movement in tubes

If you place some water in a very narrow glass tube, water will actually rise up the tube. The narrower the tube, the higher the water moves. This is called capillarity. Paper towels soak up a lot of water due to capillarity. The water creeps between the fibers of the towel. Water from the ground would seep up through house bricks, without a damp-proof layer to keep them dry.

Capillarity is a result of water molecules being attracted to the substance around them more strongly than they are attracted to other water molecules. Attraction to other water molecules is called cohesion, while attraction to other substances is called adhesion. We have already seen how powerful cohesion can be in surface tension (see page 16). However, in a tube the water molecules near the edge adhere to the glass more than to the water and are pulled upward slightly, forming a curve.

Capillarity is very important in helping water move around in plants. You can prove that water moves through plants by carrying out a simple experiment. Take a potted plant and water the soil. Wrap the pot in a plastic bag, securing the bag around the plant stem so that water cannot evaporate. Now put the whole plant, including the pot, in another, clean plastic bag. After several hours you will see condensation, in the form of water droplets on the inside of the bag. This shows that water is lost from the plant's leaves, an effect known as transpiration. But how did the water get from the roots to the leaves in the first place? There are tiny tubes inside the stem of a plant and these tubes run from the roots to the leaves. Water moves up the tubes by capillarity. They carry the water in one direction only, upward to the leaves.

! *An oak tree can lose more than 200 liters (52 gal.) of water in a day.*

EXPERIMENT

Water on the move

This experiment demonstrates capillarity in plants. You will need a white flower like a daisy or carnation, some food coloring and a beaker containing water.

1 Add a few drops of food coloring to the water in the beaker.

2 Place the cut stem of the flower in the colored water and leave it for a few hours.

3 When you return you should see that the color has been carried up to the petals, giving a colored, network appearance. If you want a two-colored flower you will need two beakers and two colors.

4 Cut the bottom part of the stem up the middle lengthways and put one half in one color of water and put the other half in the second. You will probably have to support the flower. Leave it for a few hours.

What color has the flower become?

If you want to see the tiny tubes that carry the water you could try a similar experiment.

5 Take a piece of celery and place the cut end in colored water. Leave it for an hour and then remove it.

6 Using a knife, carefully cut the bottom 2 cm or 1" off.

Can you see any tubes? The colored water should show up against the white stem, marking the position of the tubes.

How far up did the water reach in the hour in which you left it?

Key words

Capillarity attraction between water molecules and the molecules of the tube.

Hydraulic concerned with transmitting pressure through a fluid in a tube.

Pump a device that moves fluids from one place to another.

ne giant sequoias are the tallest living ees. They can reach heights of over 0 m (330 ft.). Water has to be carried om the roots to the highest branches.

The future

Huge improvements have been made in the design of ships, aircraft and cars over the last 100 years and the pace of change can be expected to continue.

The only parts of the world yet to be truly conquered by human ingenuity are the ocean depths. Strange animal forms are known to live deep in our oceans and they have evolved ways of coping with the enormous water pressure that is experienced at such great depths. As yet, humans have been unable to design a vehicle that can operate at such depths for any significant length of time, and the deepest ocean trenches are almost completely unexplored. As more and more of the limited resources on land are being used up, attention is turning toward marine resources. Machines are being designed to work in very deep water for mining and recovery of fossil fuels such as oil and gas. The environment of the very deepest oceans is too hostile for a diver, even with special gas to breathe and protective suits, so it will be necessary to design robotic vehicles. These vehicles will probably operate almost entirely on their own, with very little control from the surface. Some may use artificial intelligence to evaluate and solve problems.

Many fish like this fangtooth can survive at great depths that machines are still unable to reach. This submersible reaches depths of only 1,300 m (4,290 ft.). The deepest ocean trenches extend to 11,000 m (36,300 ft.).

The Stealth fighters are less streamlined than most other planes, but beyond 50 km (31.25 mi.) they are invisible to radar.

In the air there will be advances in aerodynamics, perhaps with new aircraft having the ability to change wing shape just like a bird. There will also be more use of composite materials and ceramics, to provide ever stronger and lighter airframes and components. While most aircraft designs use very smooth, low drag, curved surfaces, some of the latest specialized aircraft have rejected this design for a particular reason. The latest Stealth fighters do not have a typical streamlined shape, but instead have a very angular shape composed of flat, inclined surfaces that help it avoid detection by radar. A trend in passenger aircraft is toward the ability to carry more passengers. Today's jumbo jet, the Boeing 747, which can carry 400 passengers, wi be replaced by aircraft capable of carrying up to 700. One proposal by Lockheed is to join two C5 Galaxies to form a double jumbo! Another possible development will be in the realm of supersonic transports. The Concorde may be replaced by the British-designed Hotol and the American Orient Express, both of which would fly at the very top edge of the atmosphere, and could get from London to Sydney, Australia, in five hours non-stop, instead of the current 20 hours or so. These aircraft would be powered by special types of rocket engines and the planes would reach speeds of Mach 20.

At sea, humans are developing a new range of higher speed vessels to meet the demands for comfort and speed when crossing stretches of water such as the English Channel. Large catamarans, twin-hulled boats, have just entered service, and even larger ones are promised in the future. The Japanese have a plan for a vehicle that is a cross between a boat and a plane. It would be shaped like a flying wing, and travel just a few meters above the waves, relying on the surface of the sea to give added support to the air below the wing. This provides added lift, and is known as ground-effect.

The maglev train floats a few millimeters above its track, kept in place by a strong electromagnetic field. This design eliminates friction because there is no contact between the train and the track.

On land, the efficiency of motor vehicles is continuously being improved. With the increasing importance of energy conservation and pollution reduction, new designs are appearing that are powered by electricity, contain recyclable components and minimize friction by having ever more slippery shapes. There will be more use of electric transport, such as high-speed railway trains between cities, and trolleys, for short journeys.

On the human front, too, the pace of change is relentless. In surgery, composite materials and computer-aided design are helping to make replacement joints last longer. Artificial heart valves and pacemakers are already commonplace, and a successful artificial heart is probably only a few years away.

Human technology is improving all the time, and it touches our lives in thousands of different ways. But it may just be that the answers to many engineering problems have already been designed. They are just awaiting discovery among the marvels of the living world.

Glossary

airfoil wing shape that cuts through the air and produces lift.
atom smallest particle of any chemical element that can exist alone and still be that element.
buoyancy the ability to rise or float in water.
capillarity attraction between water molecules and the molecules of a tube.
composite material a material made from two or more different materials.
convergent evolution the development of a similarity between two groups of organisms that were once different.
density a measure of how closely the atoms or molecules of a substance are packed together.
drag the resistance to movement. The flow of air or water over a moving object tends to slow it down.
energy the ability to do work.
friction a force that acts between two objects that are in contact and that tends to prevent the movement of one surface over another.
gravity a force that draws two objects together.
hydraulic concerned with transmitting pressure through a fluid in a tube.
laminar flow the smooth flow of a fluid over a streamlined surface without turbulence.
lift the upward force that counteracts weight.
mammal an animal species in which the female gives birth to live young and produces milk.
marine animal an animal that lives in salt water.
mass a measure of how heavy an object is.
molecule the smallest naturally occurring particle of a substance, made from two or more atoms.
organism any living thing.
pump a device that moves fluids from one place to another.
resistance any force that slows down or opposes movement.
streamlining the design of a smooth, slippery shape that keeps drag to a minimum.
surface tension molecular force that pulls the surface of a liquid into the minimum area possible.
thrust the force that moves an object forward.
turbulent flow the rough agitated flow of a liquid over a surface.
valve a structure that allows a fluid to flow in one direction only.
vertebrate an animal that has a backbone.
volume the amount of space taken up by something.

Answers to the questions

p. 8 Float higher in salt water as salt water is more dense. To test – use a hydrometer.
p. 10 Both are containers of gas, volume can be changed, both located within the body of the structure.
p. 13 All streamlined, slippery shape for minimum resistance to flow of water, tapered at one end but rounded at the other end, all have fins.
p. 14 Designed to keep car on ground, to increase down force.
p. 16 To catch less wind and prevent capsizing, help control the boat in bad weather.
p. 17 Water droplet stays together as it is moved due to surface tension.
p. 17 Splayed legs – to reduce the pressure on the surface of the water at any one point and prevent insect falling through surface, also large area covered by legs therefore greater chance of detecting vibrations in water.
p. 21 1 – C, 2 – A, 3 – D, 4 – B
p. 24 When landing both the Concorde and many birds stall (i.e., become almost vertical in the air so hind part of bird/plane touches down first). Other aircraft land in near horizontal position.
p. 25 To reduce competition between parent plant and offspring and to colonize new areas.
p. 27 Harrier – lift from downward directed jets of air, insects beat wings in such a way as to produce a similar downward jet of air.
p. 28 Use engines to direct air flow up or down.
p. 31 For better grip as they sweat, powder increases friction.
p. 32 Wet weather – less friction between tire and road, water acting as lubricant.
p. 32 Cheetah – long, slim muscular body, long tail held out to balance when running, this helps cornering, etc. Head held in same position while running so eyes can keep focus on moving prey.
p. 33 Used in chairs, nets on aircraft carriers to stop incoming planes, racket strings in sport, bungee jumping.
p. 34 To reduce impact damage, stop heel hitting ground with too much force and causing injury.
p. 36 Longer wrench.
p. 36 Examples – nutcracker, scissors, door handle, toilet flush mechanism, wheelbarrow, pliers, wrench, car jack.
p. 37 Car trailer hitch, some camera tripods.
p. 39 Air will be compressed and so change volume and reduce the efficiency of the hydraulic system.
p. 41 The atria contract so that the ventricles fill up before they contract. Faulty heart valves would mean that blood would flow between the chambers before they were ready to contract, and there could be leakage backwards.

Index

Key words appear in **boldface** type.